Schriftenreihe aus dem Institut für Ström

Herausgeber
J. Fröhlich, S. Odenbach, K. Vogeler

Institut für Strömungsmechanik
Technische Universität Dresden
D-01062 Dresden

Band 11

Sascha Heitkam

Manipulation of liquid metal foam with electromagnetic fields: a numerical study

TUD_press_

2014

Die vorliegende Arbeit wurde am 24. März 2014 an der Fakultät Maschinenwesen der Technischen Universität Dresden als Dissertation eingereicht und am 23. Juni 2014 erfolgreich verteidigt.

This work was submitted as a PhD thesis to the Faculty of Mechanical Science and Engineering of TU Dresden on 24 March 2014 and successfully defended on 23 June 2014.

Gutachter | Reviewers
Prof. Dr.-Ing. habil. Jochen Fröhlich, TU Dresden
Prof. Dr. habil. Dominique Langevin, Université Paris Sud
Prof. Dr. habil. Reinhard Höhler, Université Pierre et Marie Curie Paris
Prof. Dr. habil. John Banhart, Helmholtz Zentrum Berlin

Bibliografische Information der Deutschen Nationalbibliothek
Die Deutsche Nationalbibliothek verzeichnet diese Publikation in der Deutschen Nationalbibliografie; detaillierte bibliografische Daten sind im Internet über http://dnb.d-nb.de abrufbar.

Bibliographic information published by the Deutsche Nationalbibliothek
The Deutsche Nationalbibliothek lists this publication in the Deutsche Nationalbibliografie; detailed bibliographic data are available in the Internet at http://dnb.d-nb.de.

ISBN 978-3-944331-92-8

© 2014 TUDpress
Verlag der Wissenschaften GmbH
Bergstr. 70 | D-01069 Dresden
Tel.: 0351/47 96 97 20 | Fax: 0351/47 96 08 19
http://www.tudpress.de

Manipulation of liquid metal foam with electromagnetic fields: a numerical study

Von der Fakultät Maschinenwesen
der Technischen Universität Dresden

zur

Erlangung des akademischen Grades
Doktor-Ingenieur (Dr.-Ing.)
angenommene

Dissertation

von

Dipl.-Ing. Sascha Heitkam

geboren am

18.12.1984 in Hoyerswerda

Tag der Einreichung: 24.03.2014
Tag der Verteidigung: 23.06.2014

Gutachter:

Prof. Dr.-Ing. habil. Jochen Fröhlich
Technische Universität Dresden

Prof. Dr. habil. Dominique Langevin
Université Paris Sud

Prof. Dr. habil. Reinhard Höhler
Université Pierre et Marie Curie Paris

Prof. Dr. habil. John Banhart
Helmholtz Zentrum Berlin

Essentially, all models are wrong,
but some are useful.

George Edward Pelham Box, British mathematician

Acknowledgements

This work would not exist without my supervisor Jochen Fröhlich. In his lectures he introduced me to scientific computation of fluid dynamics. During my thesis he kept pushing me to work, to present and to publish in an precise and meticulous way and to keep scrutinising my results. This sometimes demands a lot of patience, but I grew to learn the scientific value of this way of working. I am grateful for the considerable amount of work he investigated to do so and for many fruitful discussions.

During my stay in Paris I learned so much about foam and the fun of working with it. I would like to thank Dominique Langevin, Wiebke Drenckhan, Anniina Salonen and the whole group "liquid interfaces" at the LPS for introducing me to the scientific wonders of foam, the inspiring atmosphere and for their habit of promoting their young academics.

And especially, I want to thank Wiebke for putting me in contact with the nice and interesting foam community, which strongly encourages me to keep working in the field of foam. Also, her guidelines for the behaviour at a conference have proven most useful (and fun).

I also want to mention Tobias Kempe and Stephan Schwarz who invested considerable effort in the development of PRIME. I and several co-workers greatly profit from their efforts, owing them gratitude.

In my thesis I had several students working with me. I would like to thank them for their contribution to this work. They have brought in their skills, creativity and commitment to accomplish their given tasks and I hope they also profited from our cooperation.

An important aspect for research is the environment, one is working in. I would like to thank my colleagues in Dresden as well as in Paris for scientific and non-scientific discussions and for distraction when it was needed. And Michel, thank you for the Pfannkuchen.

An dieser Stelle möchte ich auch meinen Eltern danken, für ihre jahrelange Unterstützung, sowohl moralisch, als auch mit Rat und Tat und nicht zuletzt finanziell. Ihr habt mich motiviert, meinen eigenen Weg zu gehen und mir dabei nie Grenzen gesetzt. Und auch dem Rest meiner biologischen und angeheirateten Familie danke ich für ihre Unterstützung und für die notwendige Ablenkung.

Meinem Sohn Nemo danke ich dafür, dass er mich jeden Morgen um 5 Uhr weckt, damit ich pünktlich auf Arbeit bin und dafür, dass er mir zeigt, dass Arbeit nicht alles ist.

Aber besonders möchte ich meiner Frau danken. Dafür, dass sie oft bei ihrer eigenen wissenschaftlichen Karriere Kompromisse eingegangen ist, um mir die Zeit und Muße zu geben, mein Thema voranzubringen. Dafür, dass sie mich in allen Lebenslagen unterstützt und erträgt. Aber vor allem dafür, dass sie es in den letzten Wochen mit einem Minimum an Nörgeln geduldet hat, dass ich mich nahezu komplett dieser Arbeit gewidmet habe während sie die Familie am laufen hielt. Dankeschön!

This work was funded by the Collaborative Research Centre SFB609, by the french gouvernment via a Bourse d'Excellence Eiffel, and by the European Centre for Emerging Materials and Processes Dresden (ECEMP). Computation time was provided by the ZIH Dresden.

Graphical abstract

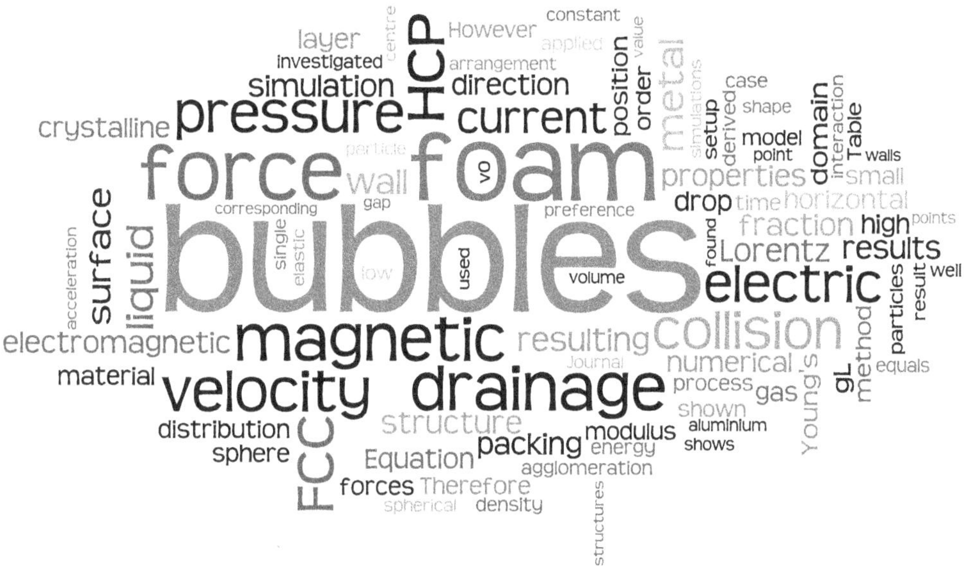

Figure 1: *Visualisation of the 100 most frequently used words in this thesis, created with Wordle (©Jonathan Feinberg 2013). Common words like 'the' are neglected.*

Abstract

Metal foam has unique mechanical and thermodynamic properties which could prove useful in many fields, such as light-weight construction and automotive engineering. However, metal foam is not yet established in engineering. One reason are the difficulties and high prices in the fabrication process. Caused by gravity-driven drainage in liquid state, inhomogeneous material distributions can occur. Also, exceeding drainage might cause bubble rupture and the generation of blow-holes. These negative effects potentially can be avoided by adding magnetic or electromagnetic fields during the generation process. In this thesis, the influence of these fields is therefore investigated by conducting phase resolving simulations.

These simulations are carried out with the in-house code PRIME. A modification of the collision modelling was necessary in order to investigate the agglomeration of bubbles within liquid metal.

Computing a static-drainage setup, the agglomeration mechanisms are investigated without the presence of electric or magnetic fields. At high drainage velocities the bubbles float. At lower drainage velocities the bubbles agglomerate in the upper part of the domain, forming close-packed crystalline structures. The experimentally well known preference of face-centred cubic ordering, over hexagonally close-packed ordering of equal-volume bubbles is reproduced numerically. Applying further simulations and experiments, an instability mechanism of hexagonally close-packed ordering is identified, resulting in the preference of face-centred cubic ordering.

In order to determine the mechanical properties of solid metal foam with low gas fraction and to state advantageous and disadvantageous features of bubble arrangements, Finite-Element simulations of solid metal foam with spherical voids are carried out and compared. A significant influence of the amount of bubble crystals on the foam mechanics is found. The type of the crystalline ordering is less important.

The influence of a horizontal magnetic field on the bubble agglomeration is investigated. The drainage resistance can be increased significantly by adding a magnetic field. The resulting structure of the bubbles is less sensitive to a magnetic field.

Combining a horizontal electric current and a perpendicular, horizontal magnetic field results in a vertical Lorentz force. This force can balance gravitation and thus, cause flotation of the bubbles. Simulating this state reveals a homogeneous bubble distribution. At the same time, curling force fields in the vicinity of each bubble induce a continuous stirring motion. Small electromagnetic field strengths do not prevent the bubbles from agglomerating, but can vary the amount of crystallized ordering and therefore, the mechanical properties of the resulting solid foam.

In conclusion, a horizontal magnetic field increases the drainage resistance, while its combination with an electric current causes homogeneous bubble distributions and can alter the foam structure and the gas fraction. The results of this thesis could help improve the industrial fabrication process of metal foam or even allow production of porous metal with user-defined gas fraction.

Zusammenfassung

Metallschaum hat einzigartige mechanische und thermodynamische Eigenschaften. Daher könnte er in vielen Bereichen, beispielsweise im Leichtbau oder Fahrzeugbau Anwendung finden. Dennoch hat sich Metallschaum im Ingenieurwesen noch nicht etabliert. Ein Grund dafür sind die Schwierigkeiten in der Herstellung. Durch die Schwerkraft kann es im flüssigen Zustand zu Drainage und damit zu inhomogenen Materialverteilungen kommen. Starke Drainage führt schließlich zu Blasenzerfall und damit zur Bildung von Lunkern. Diese negativen Effekte können möglicherweise durch den Einsatz von magnetischen und elektromagnetischen Feldern im Entstehungsprozess vermieden werden. Um das zu untersuchen, wurden in dieser Arbeit phasenaufgelöste Simulationen von der Entstehung von Metallschaum unter Einfluss elektromagnetischer Felder durchgeführt.

Für die Simulationen wurde der institutseigene Code PRIME verwendet. Ein Kollisionsmodell ist entwickelt und implementiert worden um die Interaktion von Blasen zu beschreiben.

Mithilfe der Simulation einer "Stationären Drainage"-Konfiguration wurden die Mechanismen von Blasenaufstieg und Agglomeration untersucht. Hohe abwärts gerichtete Volumenströme halten die Blasen in der Schwebe. Bei geringeren Volumenströmen sammeln sich die Blasen am Deckel der Domain und bilden dichteste Kugelpackungen. Die experimentell bekannte Bevorzugung von kubisch flächenzentrierter Packung gegenüber hexagonal dichtester Packung wurde hier numerisch bestätigt. Mithilfe gezielter numerischer und experimenteller Untersuchungen wurde ein Instabilitätsmechanismus in hexagonal dichtester Packung gefunden und damit erstmals die Bevorzugung von kubisch flächenzentrierter Packung in Blasenagglomerationen erklärt.

Um zu untersuchen, welche Blasenverteilungen in Metallschaum überhaupt günstig oder ungünstig sind, wurden Finite-Elemente Simulationen von festen Metallschäumen mit kugel-förmigen Blasen in verschieden Packungen durchgeführt und verglichen. Die Häufigkeit von dichter Kugelpackung hat einen wesentlichen Einfluss auf die mechanischen Eigenschaften des Metallschaums. Die Art der Packung ist jedoch relativ unwichtig.

Der Einfluss eines horizontalen Magnetfeldes auf die "Stationäre Drainage" wurde untersucht. Der Drainagewiderstand kann durch ein Magnetfeld wesentlich erhöht werden, nicht jedoch die resultierende Blasenpackung.

Aus der Überlagerung eines horizontalen Magnetfeldes mit einem senkrecht dazu orientierten, horizontalen Strom resultiert eine vertikal gerichtete Lorentzkraft. Diese kann die Gravitationskraft kompensieren und die Blasen in einen Schwebezustand versetzen. Die Simulation dieses Zustandes ergab eine homogene Blasenverteilung. Gleichzeitig werden rotierende Kräfte in der Nähe der Blasenoberflächen induziert, die für eine Rührbewegung sorgen. Reicht das elektromagnetische Feld nicht aus, um die Gravitation zu kompensieren, steigen die Blasen auf und sammeln sich am Deckel. Das Feld beeinflusst hier jedoch die Häufigkeit dichter Kugelpackung und damit die mechanischen Eigenschaften des resultierenden Metallschaums.

Zusammengefasst erhöht ein horizontales Magnetfeld den Drainagewiderstand des Schaums, während ein elektromagnetisches Feld die Blasen schweben lassen und homogen verteilen oder ihre Agglomeration beeinflussen kann. Die Ergebnisse dieser Arbeit haben das Potential, die industrielle Herstellung von Metallschaum zu verbessern. Durch das gezielte Verteilen der Blasen im Schwebezustand wäre es sogar möglich völlig neue poröse Materialien mit beliebigem Gasgehalt zu produzieren.

Contents

1 Introduction   1
   1.1 The challenge of working with foam   1
   1.2 Generation and disintegration of foam   1
   1.3 Metal foam   4
   1.4 Metal foam fabrication   5
   1.5 Measurement versus simulation   6
   1.6 Material properties   7

2 The multiphase code   9
   2.1 Capabilities of the code   9
   2.2 The fluid solver   9
   2.3 Immersed-Boundary method   10
   2.4 Simulation of magnetohydrodynamics   11

3 Collision modelling   15
   3.1 Motivation   15
   3.2 Basic collision modelling   15
   3.3 Other types of collision modelling   17
   3.4 Advanced collision model   18
   3.5 Testing and experimental validation   30
   3.6 Conclusions   34

4 Rise and agglomeration of bubbles   37
   4.1 Motivation   37
   4.2 Rise of a single bubble   37
   4.3 Static drainage   40
   4.4 Agglomerated regime   41
   4.5 Floating regime   42
   4.6 Structure formation   44
   4.7 Conclusions   45

5 Bubble crystals   49
   5.1 Motivation   49
   5.2 Setup   51
   5.3 Quantification of crystalline order   52

5.4 Hydrodynamic effects in bubble agglomeration 54
5.5 Mechanical stability 54
5.6 Experimental reproduction 58
5.7 Conclusions 59

6 Mechanical properties of solid foam   63
6.1 Motivation 63
6.2 Definition of sphere structures 64
6.3 Homogenisation and elastic properties 66
6.4 The Finite-Element method 67
6.5 Validation 67
6.6 General comparison of Young's moduli 70
6.7 Comparison of cubic structures 70
6.8 Comparison of hexagonal structures 70
6.9 Conclusions 75

7 Manipulation of foam with magnetic fields   77
7.1 Motivation 77
7.2 Setup 78
7.3 Results 79
7.4 Conclusions 84

8 Manipulation of foam with electromagnetic fields   87
8.1 Motivation 87
8.2 Physical background 87
8.3 Setup 89
8.4 Behaviour of a single bubble 90
8.5 Interaction between a pair of bubbles 94
8.6 Manipulation of bubble clusters 96
8.7 Artificial bubble distribution 99
8.8 Conclusions 101

9 Influence on foam structure   105
9.1 Motivation 105
9.2 Setup 105
9.3 Results 106
9.4 Conclusions 110

10 Concluding remarks   113

11 Appendix   129
11.1 Symbols 129
11.2 The author's publications 133
11.3 Student theses 136

1 Introduction

1.1 The challenge of working with foam

In our daily life we are surrounded by foam. A lot of cosmetic and food products contain or generate foam. Also, in most thermal insulations of buildings or vehicles, foam is applied. One can say, that foam plays an important role in our daily life. Nevertheless, many processes in the generation and disintegration of foam are not yet well understood. The reason might be, that foam is an extremely complex matter, combining many chemical and physical aspects that are awkwardly difficult to understand in their manifold interaction. At the same time, foam covers processes on many length-scales from the rheological behaviour of big clusters of bubbles down to the adsorption and desorption of single molecules. Also time-scales vary between the sudden rupture of a liquid film up to the lifetime of highly stable food products. For these reasons, foam is a very challenging, but also a very interesting field of investigation. Foam, especially aqueous foam is usually cheap, everybody can afford to do foam experiments. With very easy experiments one can already experience very complex mechanisms. The art is, to design experiments that allow for separation of a limited number of effects, in order to have a chance to understand the individual effects and their interaction. Numerical simulations are predestined to separate a number of effects by just accounting or not accounting for them in the set of equations considered. Therefore, numerical simulations should play a more important role in the investigation of foam, even though they are not yet able to cover foam in its whole variety.

1.2 Generation and disintegration of foam

The book "Foams: structure and dynamics" gives a valuable overview about foams and is the basis of this section [1]. Foam is a mixture of dispersed gas bubbles and a continuous liquid. It also contains some surface active agents (surfactants) that are adsorbed in the gas-liquid interface and stabilise it. Several classes of surfactants exist, using different stabilisation mechanisms. Foam is usually generated by adding bubbles to a liquid, contaminated with surfactants. There are various ways to do so. For example, one can blow gas into a fluid, one can shake it or one can add gas-releasing substances. Driven by gravity these bubbles rise until their path is blocked by an obstacle or by the surface of the liquid. Consecutive bubbles arrive and the bubbles agglomerate. Foam is created. When the foam consist of close-packed spheres, the liquid content equals commonly around 36 % if the foam is disordered, while it equals 26 % if equal-volume bubbles are arranged on crystalline lattices. These sphere packings are called the "wet limit" and foams with liquid fraction just above this limit are called "wet foams". Due to their buoyancy all bubbles push upward, displacing the interstitial liquid. The displaced liquid moves downward through the agglomerated bubbles, which is called "drainage". By reducing its liquid fraction the wet foam transforms into dry foam, corresponding to less than 10 % liquid fraction. At the same time, bubbles of the foam lose their spherical shape and thus, increase their surface area and surface energy, respectively. This increase in energy results in a quasi-osmotic pressure that tries to suck in liquid, and thus, counteracts gravitation [2, 3]. Drainage will stop, when the gravitationally induced pressure gradient is completely counteracted by the osmotic pressure gradient. Consequently, foam cannot contain infinitely small amounts of liquid but approaches asymptotically an equilibrium state with a finite amount of liquid content. Figure 1.1 shows the structure of dry soap foam.

From a geometrical point, three-dimensional, perfect dry foam consists of lamellas,

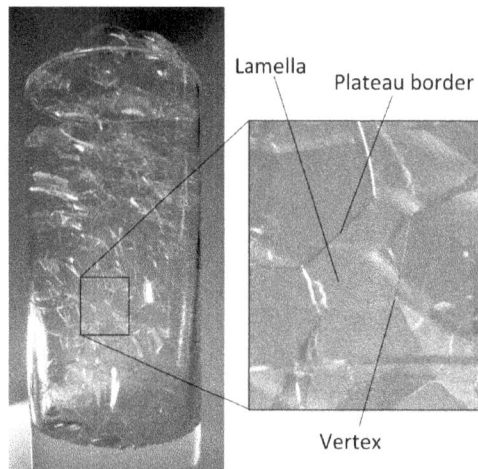

Figure 1.1: Picture of dry soap foam, showing lamellas, Plateau borders and vertexes.

Plateau borders and vertices. Lamellas are thin layers of liquid, separating two bubbles. Plateau borders are straight or slightly curved lines that connect three lamellas under an angle of 120°. Four Plateau borders meet in a vertex under the tetrahedral angle of 109.5°. These restrictions are described by the Plateau's laws [1]. The terms "lamella", "Plateau border" and "Vertex" will also be applied to wet foam throughout this thesis.

Realistic foam is not stable in its equilibrium state. Foam coarsening takes place, which means the bubble size distribution is shifted towards larger bubbles until the foam disintegrates. Figure 1.2 shows beer foam in three different states. Right after pouring, it is nicely homogeneous. After 30 seconds, coarsening results in signifi-

Figure 1.2: Beer foam in three different stages.

cantly larger bubbles. Due to drainage, evaporation and bubble rupture the foam is dissolved on top. After 5 minutes most of the foam is gone. The remainings of some large bubbles are still visible.

Coarsening is caused by two effects, coalescence and diffusion. The diffusion effect is called Ostwald ripening for spherical bubbles and von Neumann-Mullins coarsening in dry foam [1]. Both coarsening mechanisms, coalescence and diffusion, are illustrated in a non-physical way in Figure 1.3. Coalescence describes the combination

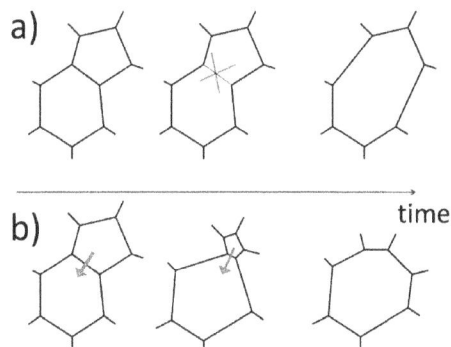

Figure 1.3: Illustration of the two mechanism of foam coarsening. Coalescence (a) is the combination of two bubbles by film rupture, while von Neumann-Mullins coarsening (b) is the emptying of bubbles by diffusion.

of two bubbles by destruction of the separating lamella. If a lamella becomes very thin, its two liquid-gas interfaces are separated by the disjoining pressure, a repelling force between the surfactants in each interface. Eventually, this balance can become locally distorted, leading to rupture of the lamella [1, 4]. Coalescence can be reduced by adding sufficient amounts of surface active agents with high disjoining pressure, as it is done in whipped cream. The other coarsening process, diffusion, causes the growth of large bubbles and the disappearing of small ones. The reason is, that smaller bubbles have a higher pressure than bigger bubbles, resulting in diffusion of the gas from small to big bubbles. This process can be minimised by applying gases with low solubility in the applied liquid.

Both aspects of foam coarsening can be reduced very effectively by keeping a high content of liquid and thus, thick lamellas. Therefore, foam can be stabilised to a large extent by adding particles or droplets to the fluid that hinder drainage [5].

1.3 Metal foam

Metal foam is a foam made from gas bubbles enclosed in a metal, such as copper, zinc aluminium or steel. The gas fraction equals between 80 and 95 %. Therefore, metal foam has a very low density, compared to pure metal. After solidification, metal foam has unique properties, such as high stiffness and high stability, compared to other materials with comparable density. These properties are particularly useful in light-weight construction. To benefit from them, metal foam is applied in some kind of sandwich-design, shown in Figure 1.4, taken from [6]. In that way, the cover layers absorb most of the mechanic load, while the foam maintains the distance between the layers and avoids buckling.

The Fraunhofer IWU in Chemnitz, Germany, developed a slide for a HSC (high

Figure 1.4: Picture of metal foam in sandwich-design [6].

speed cutting) milling machine in this sandwich-design. In that way, they managed to reduce the slide weight by 28 %. Since the slide is an element that has to be accelerated and decellerated permanently in the production process, its lower weight allows for a faster production [7].

Another big advantage of metal foam is its high energy absorption at plastic deformation. This property qualifies for crash absorbers, e.g. in automotive engineering. In case of a crash, a certain crumple zone size is available. While deforming the crumple zone, the kinetic energy of the vehicle has to be absorbed. This can be done very effectively by metal foam. The first mass-produced aluminium foam device is a small bolt, fixing the separation net of the AUDI Q7 [8]. In case of a crash, this net has to catch the baggage. High stability and high energy absorption in crucial for this element. Another automotive application of metal foam is the frame of the Ferrari Spider430 [8] which is filled with metal foam in order to increase its stability and its crash absorbing properties.

Other useful characteristics of metal foam are high damping of oscillations and sound, low thermal conductivity and fire resistance [9, 10].

1.4 Metal foam fabrication

Despite these unique properties, up to now metal foam is only used in some rare, highly specific applications. The examples, mentioned above, also show one big drawback of metal foam. It is expensive. Thus, it is only used in expensive applications, such as luxury cars. Another problem is the lack of standardized material parameters or testing methods [8]. The high price results, among others, from the difficulties in controlling or even predicting the generation process. Another problem is to achieve reproducibility of the production process, which is necessary to guarantee material properties to the customer. Therefore, expensive material testing methods have to be applied.

Several manufacturing routes, such as external gas blowing or gas releasing blowing agents [11–13] exist. However, they all suffer from drainage and foam coarsening.

At some stage of the production process, distributed bubbles in liquid metal exist. Due to gravity, drainage removes more and more metal from the foam, as described above. If the local metal fraction becomes too small, coarsening of the foam cannot be prevented. Coalescence will take place and a coarse bubble structure or even blow-holes might result, degrading the mechanical properties of the resulting element. Hence, the control of drainage in liquid metal foam is desired. Applying zero-gravity experiments Banhart and co-workers showed, that the absence of drainage creates a more homogeneous foam [14–17]. Since zero-gravity is not available for industrial processing, other methods are necessary to suppress drainage. Figure 1.5 shows an example for favourable (a) and unfavourable (b) metal foam. Both samples result from the work of Brunke et al. [18–20]. Favourable metal foam consists of bubbles with a well controlled, not too large diameter and sufficient amount of metal. The lamellas between the bubbles are faultless. Even in this favourable state, some small blow-holes exist. But in unfavourable metal foam, coarsening has significantly increased the bubble size. The stress distribution in this type of foam would be very inhomogeneous. This material is much more likely to collapse in a practical application.

In the present thesis, the influence of magnetic fields on the drainage in metal foam is investigated. It is well known that motion of conducting liquids can be damped by a transversal magnetic fields [21–26], but the influence of such a field on drainage in metal foam has not yet been analysed.

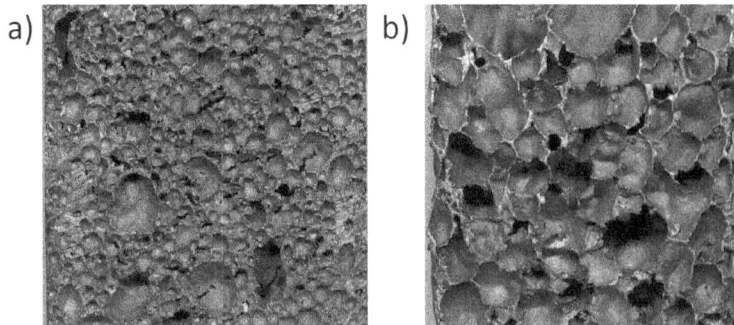

Figure 1.5: Picture of metal foam in favourable, (a), and unfavourable, (b), condition, resulting from the work of Brunke [18].

An even more challenging goal is to prevent the bubbles from rising in the first place and to distribute them equally or even artificially in the container. Leenov and Kolin [27] have shown that a combination of electric and magnetic field causes Lorentz forces that can accelerate electrically insulating particles. Motivated by these results, the influence of combinations of magnetic and electric fields on the bubble distribution, on drainage and on resulting foam properties will be investigated as well.

1.5 Measurement versus simulation

Measurments of physical quantities of aqueous foam are very challenging and there are only a few commonly used measurement techniques. The most obvious method is the optical observation. However, foam has often a very complex structure of random nature. A penetrating light beam is multiple times refracted and reflected, disturbing optical signals significantly. Direct optical measurements of the bubbles can therefore only be applied to the three outmost layers of bubbles [28]. Electrical resistance measurement was found to reveal the local liquid fraction [29]. Much effort is spent to measure the rheological properties of different types of foam, e.g. in adapted rheometers [30–32]. On a smaller scale, the influence of surfactants on the gas-liquid surface is characterized in a dynamic approach, e.g. by using the pendent-drop method. This method is widely used in order to evaluate surfactants. One can even order commercial tools [33]. An alternative is the investigation of capillary wave propagation on an air-water surface [34]. In addition to these established techniques, there is a wide range of trial stage experiments researchers came up with, in order to understand certain aspects of foam. However, most of these techniques only give limited insights to the complex interaction of the mechanisms playing a role in realistic foam. Information on velocity, pressure or concentration within the foam are rarely available.

Liquid metal foam is even more difficult to measure, since it is opaque, hot and not very stable. X-ray tomography has given some interesting insights into the evolution of metal foam in the generation process [17–20, 36]. It allows to quantify the amount of liquid, the drainage and foam coarsening. Unfortunately, to penetrate the metal and achieve high quality data, high energy radiation is necessary, which is expensive and difficult to handle.

Due to these difficulties, in this work, multi-phase flow simulations are applied. They allow for insights into the bubble agglomeration, the resulting flow and pressure field of the continuous fluid. They also allow to switch on and off effects like gravity, enabling to separate their influences. However, many principle effects like adsorption and desorption kinetics of surfactants at the interfaces [37–39] or surface viscosity [40–42] are not sufficiently understood and can therefore be included only to a very limited extent. Other effects, like flow in the thin foam lamellas [4, 43, 44], are well understood but nevertheless cannot be covered by the simulation due to numerical restrictions of the applied method. For these reasons, the numerical simulation of foam also is very challenging, drawing only an incomplete picture of the processes, taking place in foam generation. Nevertheless, the results of the present simulations will broaden the knowledge about the processes in foam. General influences and dependencies on parameters are supposed to be valid, even though the simulation is lacking some physical and chemical aspects.

1.6 Material properties

The goal of this thesis is the investigation of the agglomeration process of bubbles in a metal foam in the liquid state. Therefore, several liquid metals such as copper, iron, aluminium or zinc can be taken into account. However, practical applications often use aluminium foam [8] because aluminium is light, foamable and has a high specific stability.

Measuring the exact material parameters of liquid aluminium is very challenging, because of its high temperature and the risk of impurifications. Despite these uncertainties, one set of material parameters was extracted from literature [45, 46] and is used throughout this thesis, without scrutinising it. Also, the viscosity strongly depends on the temperature [45], which is not taken into account in this thesis. In liquid state, aluminium has, compared to water and air, a high surface tension, a high density and a low kinematic viscosity [45, 46]. Aluminium also has a high electric conductivity and a negligible interaction with magnetic fields. Thus, the interaction with electromagnetic fields can be modelled rather easily, as described in Section 2.4 below. The applied material parameters of gas bubbles in liquid aluminium are given in Table 1.1.

They have been used for all computations presented in this thesis, except for the computation of the collision experiment in Chapter 3, where water was considered. For water, the applied material properties are given in Table 1.2.

Aluminium foams can be made more stable by adding surfactants, such as aluminium oxide or silicon carbide particles [47]. These particles will agglomerate at the gas-metal interface, causing a no-slip condition [48] and stabilizing the interface. Their influence on the material properties will not be taken into account by the

Table 1.1: *Material parameters and constants for the simulations of bubbles in liquid aluminium, conducted in this thesis.*

Quantity	Symbol	Value
Fluid density	ρ_f	$2400\,\mathrm{kg/m^3}$
Fluid viscosity	ν_f	$5 \times 10^{-7}\,\mathrm{m^2/s}$
Surface tension	σ	$1\,\mathrm{N/m}$
Electrical conductivity	σ_el	$5 \times 10^5\,\mathrm{S/m}$
Gravitational acceleration	g	$9.81\,\mathrm{m/s^2}$
Gas density	ρ_b	$1.2\,\mathrm{kg/m^3}$

Table 1.2: *Material parameters and constants for the simulations of bubbles in water, conducted in this thesis.*

Quantity	Symbol	Value
Fluid density	ρ_f	$1000\,\mathrm{kg/m^3}$
Fluid viscosity	ν_f	$1 \times 10^{-6}\,\mathrm{m^2/s}$
Surface tension	σ	$0.03\,\mathrm{N/m}$
Gravitational acceleration	g	$9.81\,\mathrm{m/s^2}$
Gas density	ρ_b	$1.2\,\mathrm{kg/m^3}$

present investigations. Only the no-slip condition at the interfaces will be adopted, as described in Section 2.3.

2 The multiphase code

2.1 Capabilities of the code

In this work, the computations of fluid dynamics have been performed with the multiphase code "Phase-resolving simulation environment" (PRIME) [49–54]. It has been developed at the Institute of Fluid Mechanics, Technical University Dresden since 2006. The scope of PRIME is to simulate dispersed particles in a continuous, homogeneous fluid. The geometry of the particles and the interaction between particles and fluid is fully resolved. Interaction between particles and of particles with walls is realized by collision modelling. PRIME can deal with more than 40 000 mobile particles [55] by accepting a slightly unsharp particle surface. In the present thesis, the particles are supposed to be bubbles. By solving additional equations, PRIME can also compute magnetohydrodynamical flows [25, 54, 56] and scalar transport.

The author has only a small share in this development, mainly focused on the collision modelling as described in Chapter 3. Thus, the purpose of the present chapter is to document the applied method.

2.2 The fluid solver

The core of the code PRIME is a highly effective solver of the unsteady, incompressible Navier-Stokes-Equations [57], discretised on a staggered Cartesian grid

$$
\begin{aligned}
\frac{\partial \mathbf{u}}{\partial t} + (\mathbf{u} \cdot \nabla)\mathbf{u} &= -\frac{p}{\rho_\mathrm{f}} + \nu_\mathrm{f}\nabla^2\mathbf{u} + \mathbf{f} + \mathbf{f}_\mathrm{IBM} \\
\nabla \cdot \mathbf{u} &= 0.
\end{aligned}
\tag{2.1}
$$

Solving this differential equation provides the time-dependent fluid velocity $\mathbf{u}$ and the corresponding pressure p for a flow including volumetric forces $\mathbf{f}$ and the force $\mathbf{f}_\mathrm{IBM}$ resulting from the interaction with particles. For time advancement a low-storage three-step Runge-Kutta-scheme with implicit treatment of the viscous terms is used [53]. Turbulence modelling is available with PRIME, but is not required in the present work. Instead, the spatial and temporal resolution is chosen high enough to resolve the smallest vortices, i.e. a direct numerical simulation (DNS) [58] is performed.

2.3   Immersed-Boundary method

The code PRIME is capable of computing the movement of light or heavy spherical particles by fully resolving the geometry of the particles and the interaction between a particles surface and the surrounding fluid. This is done by applying the Immersed-Boundary method (IBM) [59, 60]. The surface is represented by Lagrangian marker points, as shown in Figure 2.1 for a spherical particle.

At these points $\mathbf{x}_L$, forces $\mathbf{F}_{IBM}$

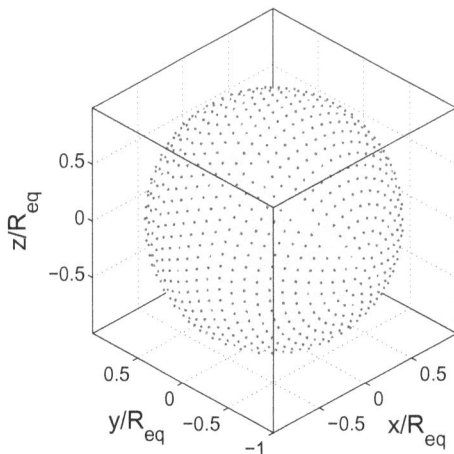

Figure 2.1: Distribution of 1258 Lagrangian marker points on a particles surface.

$$\mathbf{F}_{IBM} = \int_{\Re^3} \delta(\mathbf{x}_L) \mathbf{f}_{IBM} dV \qquad (2.2)$$

are added to the fluid, distributed by a delta-function δ. In that way, the preselected fluid boundary condition is imposed at the surface [59]. Since action equals reaction, a counteracting force acts on the particle surface. The sum of all forces $\mathbf{F}_{IBM}$ all its n_L Lagrangian marker points accelerates the particle. By solving an equation of motion (2.3) for the particle, its translation and rotation are derived

$$m_b \frac{d\mathbf{u}_c}{dt} = -\rho_f \sum_{i=1}^{n_L} \mathbf{F}_{IBM} + V_b(\rho_b - \rho_f)\mathbf{g} + \rho_f \frac{d}{dt} \int_{\Omega_b} \mathbf{u}\, dV + \mathbf{F}_{coll}$$

$$\frac{m_b D^2}{10} \frac{d\boldsymbol{\omega}_b}{dt} = -\rho_f \sum_{i=1}^{n_L} (\mathbf{x}_L - \mathbf{x}_c) \times \mathbf{F}_{IBM} + \rho_f \frac{d}{dt} \int_{\Omega_b} (\mathbf{x} - \mathbf{x}_c) \times \mathbf{u}\, dV, \quad (2.3)$$

with m_b the particle mass, D the particle diameter, V_b the volume of the domain Ω_b representing the particle, $\mathbf{x}_c$ the particle centre position, $\mathbf{u}_c$ the particle centre velocity, $\mathbf{F}_{coll}$ the collision force explained in Chapter 3, $\boldsymbol{\omega}_b$ the rotational velocity of the particle, t the time, and $\mathbf{x}$ the position in space. Detailed information about the IBM for heavy particles is given by Uhlmann [60]. The method of Uhlmann is not capable of simulating particles with a density ration between particle and fluid $\rho_\mathrm{b}/\rho_\mathrm{f}$ smaller than 1.3. Kempe and Fröhlich have extended this method by direct integration of the share of the interaction force that is consumed to accelerate the unphysical fluid inside the particle, using a second-order cut-cell approach [52, 53]. This step allows for density ratios down to 0.3. However, for the computation of foam, density ratios of about 0.001 have to be computed. If a particle is very light, e.g. if the density ratio is smaller than 0.3, the direct computation of Equation (2.3) becomes unstable. Schwarz and Fröhlich [54, 56] developed the Virtual-Mass method, scattering the fluid action on the particle in time and thus, smoothing its trajectory. With this method, it is possible to compute very light particles down to a density ratio of below 0.001. Consequently, PRIME is able to investigate the movement of particles as light as bubbles. In terms of the fluid boundary condition at the bubble surface a no-slip and a free-slip boundary condition can be chosen. Conditions for no-slip are given in Equation (2.4)-(2.5),

$$(\mathbf{u} - \dot{\mathbf{x}}_\mathrm{c} - \omega_\mathrm{b} \times (\mathbf{x} - \mathbf{x}_\mathrm{c})) \cdot \mathbf{e}_\mathrm{n} \;\; = \;\; 0 \tag{2.4}$$
$$(\mathbf{u} - \dot{\mathbf{x}}_\mathrm{c} - \omega_\mathrm{b} \times (\mathbf{x} - \mathbf{x}_\mathrm{c})) \cdot (\mathbf{1} - \mathbf{e}_\mathrm{n} \circ \mathbf{e}_\mathrm{n}) \;\; = \;\; 0 \tag{2.5}$$
$$\tag{2.6}$$

with $\mathbf{e}_\mathrm{n}$ the normal vector of the surface element. In both cases, the normal component of the velocity of the fluid relative to the respective part of the surface has to be zero since the fluid must not penetrate the bubble. This is described by Equation (2.4). The tangential boundary condition depends on the character of the interface [48]. For a no-slip condition the surface of the bubble has to be rigidified, i.e. it must not yield to shear forces. In that case, the tangential share of the fluid velocity relative to the surface has to be zero (see Equation (2.5)). In case of a free slip condition the bubble surface will yield to any shear forces. In the present work, bubbles in liquid metal foam are investigated. To generate metal foam, surface active agents, e.g. silicon carbide particles, are added that accumulate in the liquid-gas interface and therefore rigidify the bubble surface. Also, liquid aluminium often includes aluminium oxides from oxidation at an opening. This aluminium oxide also rigidifies the surface. Therefore, a no-slip boundary condition is assumed throughout the whole thesis, referring to Equation (2.4) and (2.5).

2.4 Simulation of magnetohydrodynamics

The code PRIME also offers the possibility to include electric and magnetic fields into a simulation [54, 61]. The magnetic susceptibility of gas and liquid aluminium are both close to zero. Therefore, neither gas nor liquid aluminium become magnetized by an external magnetic field and thus, the phase boundary between gas and aluminium does not influence an external magnetic field $\mathbf{B}$ [21]. It also is not influenced by the movement of the liquid aluminium, because in the considered cases the magnetic Reynolds number $R_m = \mu_0 \sigma_{el} u D$, involving the magnetic constant μ_0 and the electrical conductivity σ_{el}, is below 0.001. Thus, magnetic advection is negligible [21]. Consequently, the magnetic field only depends on its boundary conditions and is not influenced by the simulation. It usually is chosen to be constant in space and time.

However, including an electric field is much more difficult. The electric conductivity of bubbles is much smaller, compared to the conductivity of liquid aluminium. Therefore, the bubbles represent electric insulators and thus, strongly influence the electric current $\mathbf{j}$. Also, the induction term $\mathbf{u} \times \mathbf{B}$ in Equation (2.9) is not negligible. Therefore, an additional Laplacian differential equation for the correction $\Phi_{el,cor}$ of the electric potential Φ_{el} has to be solved in each time step [61, 62]:

$$\mathbf{j}_{pre} = \alpha_{el}\sigma_{el}(\mathbf{u} \times \mathbf{B} - \nabla\Phi_{el}) \qquad (2.7)$$

$$\nabla^2\Phi_{el,cor} = \nabla \cdot \mathbf{j}_{pre} \qquad (2.8)$$

$$\mathbf{j} = \alpha_{el}\sigma_{el}\left[\mathbf{u} \times \mathbf{B} - \nabla(\Phi_{el} + \Phi_{el,cor})\right] \qquad (2.9)$$

The influence of the bubbles on the electric current is included in Equation (2.9) by the phase indicator α_{el}, which is 0 inside and 1 outside of a bubble. At the surface of the bubbles, it is computed with a second-order cut-cell approach [54].

Subsequently, the Lorentz force $\mathbf{f}_L$, acting on the fluid is computed by

$$\mathbf{f}_L = \mathbf{j} \times \mathbf{B} \qquad (2.10)$$

In Figure 2.2 the electric current around a single bubble is visualized. The electric current goes from left to right. Obviously, the zero-conductivity boundary condition

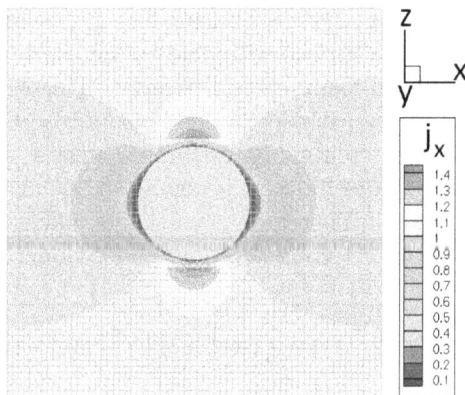

Figure 2.2: Horizontal component of the electric current around a single, insulating sphere (grey). The current goes from left to right. Red lines mark the utilized grid. Colour represents the x−component of the electric current.

for the bubble is achieved.

The boundary conditions for the electric current at the boundaries of the domain correspond to the assumed material of the domain walls. If the walls have a significantly higher electric conductivity than the liquid, the tangential component of the electric current at the walls is zero. If the walls have a significant lower conductivity, the normal component is zero. In PRIME, one can set the normal and the tangential components of the electric current or leave them variable. In this thesis, in Chapter 8 and 9 an electric current in $x-$ direction is imposed. The corresponding boundary conditions are given in Table 2.1, where $\mathbf{e}_n$ is the unit vector normal to the wall and $\mathbf{e}_{x,y,z}$ the unit vector in x-, y-, and z-direction, respectively.

Much more information on the implementation and validation of the applied mag-

Table 2.1: *Wall boundary conditions for the simulation of an electric current* $j_0\mathbf{e}_x$ *in* $x-$ *direction.*

Wall orientation	Wall conductivity	Normal current	Tangential current
$\mathbf{e}_n \parallel \mathbf{e}_x$	conductive	j_0	variable
$\mathbf{e}_n \parallel \mathbf{e}_y$	insulating	0	variable
$\mathbf{e}_n \parallel \mathbf{e}_z$	insulating	0	variable

netohydrodynamic solver can be found in the thesis of Schwarz [62].

3 Collision modelling

3.1 Motivation

When a bubble approaches an obstacle, e.g. a wall or another bubble, it deforms. This deformation corresponds to a collision force $\mathbf{F}_{\mathrm{coll}}$, acting between bubble and obstacle. However, in the employed version of PRIME bubbles are restricted to maintain a spherical shape. Since the deformation is not resolved by the simulation, it cannot automatically produce the collision forces acting between the bubble and an obstacle. At the same time, if the bubble is situated close to an obstacle, the flow and the pressure within the gap between them is underresolved since an Eulerian grid with constant step size is employed. This results in a significant underestimation of the forces acting across the gap on the bubble. Thus, a bubble approaching a wall or another bubble would overlap with it if no precautions are taken. Consequently, collision forces $\mathbf{F}_{\mathrm{coll}}$ have to be added to the bubbles equation of motion (2.3) in order to obtain the correct bubble trajectory.

In this chapter, a basic and an advanced collision model for the interaction force between a small bubble and a wall or another bubble is derived. In most of the simulations, presented in this thesis, the basic model is used because it yields sufficiently accurate results, performs very robust and requires low computational effort. However, in some cases, when the behaviour of the bubble cluster is expected to be dominated to a large extent by the bubble interaction, the simplicity of the basic model might result in unphysical results. Therefore, for the complex manipulation of foam structure in Chapter 9 the advanced collision model is used.

3.2 Basic collision modelling

In the literature, soft sphere interaction is often modelled with a spring model [63, 64]

$$\mathbf{F}_{\mathrm{coll}} = C_{\mathrm{coll}} \max[\Delta; 0]\mathbf{e}_{\mathrm{n}}, \tag{3.1}$$

where $\mathbf{e}_{\mathrm{n}}$ the unit vector pointing from the obstacle to the bubble centre, and C_{coll} an arbitrary constant. The distance deficit Δ can be derived from the centre position $\mathbf{x}_{\mathrm{c}}$ of the bubble, its diameter D, and Δ_0 which is the distance, at which the collision force starts to act

$$\Delta = 1/2 \left(D - |\mathbf{x}_{c,2} - \mathbf{x}_{c,1}| + \Delta_0 \right) \tag{3.2}$$

$$\Delta = D/2 - \mathbf{x}_c \cdot \mathbf{e}_n + \Delta_0. \tag{3.3}$$

Equation (3.2) describes the collision between two bubbles while Equation (3.3) is valid for the collision of a bubble with a plane wall. The basic geometry is sketched in Figure 3.1. If Δ_0 is too small or even equals zero, the collision force might not be

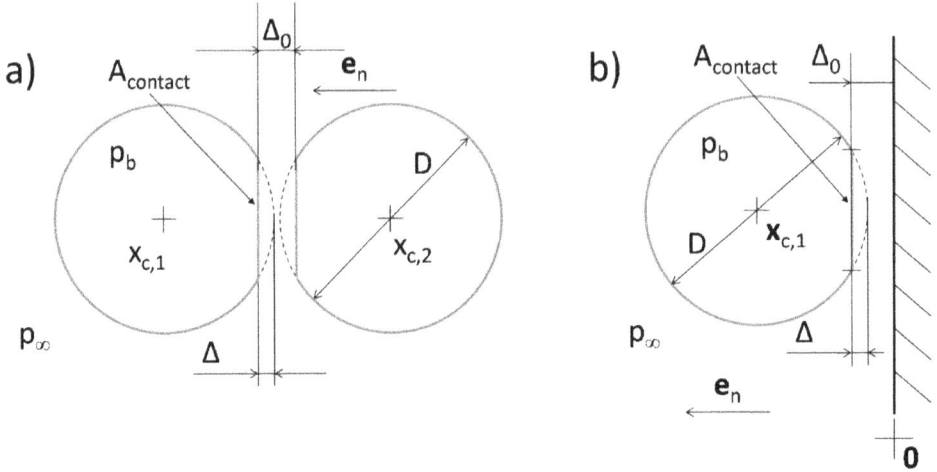

Figure 3.1: Sketch of the modelled deformation of a bubble interacting with another bubble (a) or with a wall (b). The bubble is shaped like a spherical hemisphere. The distance deficit Δ is cut away.

strong enough to deflect the bubble, before an overlap between the spherical bubble and the obstacle appears. This overlap has to be prevented, because it causes unphysical divergences in the fluid and therefore the incompressible computation crashes. To avoid this, one can either switch off the overlapping part of the surface of both bubbles by setting to zero the force $\mathbf{F}_{\text{IBM}}$ at the respective Lagrangian marker points, which avoids strong divergences in the flow field. An alternative is to choose a value for Δ_0 which is sufficiently large to avoid overlapping. In this work, the latter is done. The value for Δ_0 is set to equal two numerical grid spacings Δs which corresponds to values of approximately $0.1D$ with the used choice of parameters applied in this thesis. Higher values of Δ_0 can not be used in the recent implementation of PRIME due to the communication scheme in the parallelisation. However, $\Delta_0 = 2\Delta s$ has proven sufficient to avoid overlapping in the simulations below.

The stiffness of the bubble C_{coll} was derived from the overpressure $p_b = 4\sigma/D$ within the bubble and from the geometry, sketched in Figure 3.1. This yields a collision force

$$
\begin{aligned}
\mathbf{F}_{\text{coll}} &= p_{\text{b}} A_{\text{contact}} \mathbf{e}_{\text{n}} \\
&= \frac{4\sigma}{D} \pi (\sqrt{D\Delta - \Delta^2})^2 \mathbf{e}_{\text{n}} \\
&\approx 4\pi\sigma\Delta\mathbf{e}_{\text{n}}.
\end{aligned}
\tag{3.4}
$$

The linear approximation in Equation (3.4) is valid for small deformations $\Delta << D$. In most part of this work, this spring model with $C_{\text{coll}} = 4\pi\sigma$ is applied, even though the author is aware of the simplicity of this approach. It performs robust, demands only very low computational effort and yields good results. Especially in Chapter 4 where the recomputation of the preference of FCC over HCP structure in bubble agglomeration with the basic collision model succeeds, it proves its applicability. However, before focusing on the manipulation of the agglomeration structure in Chapter 9, the author decided to include a more sophisticated collision model.

3.3   Other types of collision modelling

For solid, heavy particles the deformation of the particle during collision is localized and well investigated [65, 66]. Also the flow and the resulting forces can be evaluated using Stokes-flow approximations [67–69]. Thus, simple collision models for this class of particles exist [70]. From a numerical point of view the small contact time is critical. It should be discretised by at least 10 numerical time steps [71], in order to resolve the collision process sufficiently. Kempe et al. [71, 72] overcome this problem by restricting the loss of kinetic energy of the particle during collision and the desired collision time and calculated and set the corresponding elastic properties of the particle just before the collision takes place.

For light, deformable particles these approaches are not applicable for several reasons. Due to the high deformation during collision simple deformation models lose validity. Also, the flow and pressure field in the gap between bubble and obstacle cannot be prescribed without knowing the shape of the bubble. At the same time, restriction of kinetic energy does not yield sufficient results, since the kinetic energy of light particles is low and thus, unimportant for the collision process.

Another approach toward collision modelling of soft particles or bubbles is the unsteady simulation of the collision zone. Hendrix et al. [73] found an elegant way to measure the lamella thickness of a bubble during the collision process using optical interference. Applying a numerical simulation of the flow field and simultaneously, of the surface shape, they found interesting numerical results on the behaviour of a lamella during collision [74–77]. The essential drawback of this method is, that it needs substantial computational effort to calculate a single collision. Thus, it is not feasible to simulate in that way systems of hundreds of interacting particles.

In the present work, an advanced collision model is developed, taking the physics of the collision zone into account. An explicit forcing scheme, similar to Equation (3.4) is derived, which is easy to implement, needs low computational effort and relies on well known, macroscopic parameters. It includes elastic and dissipative contributions to the total collision force.

3.4 Advanced collision model

3.4.1 Basic concept

The equation of motion for a particle or bubble reads:

$$\rho_b V_b \ddot{\mathbf{x}}_c = \mathbf{F}_{fluid} + \mathbf{F}_{coll} \tag{3.5}$$

The fluid force $\mathbf{F}_{fluid}$, is the force that the surrounding fluid exerts on the bubble. It includes buoyancy and could result from a phase-resolving simulation [53] as it is done in this work. It could also result from the Basset-Boussinesq-Oseen approach [78–80] calculating the fluid forces on a particle without resolving the particle-fluid interactions. In case of collision, additional forces $\mathbf{F}_{coll}$ act on the bubble. Since the bubble density ρ_b is much smaller than the density of the surrounding fluid ρ_f, the left-hand side of Equation (3.5) is negligible. Thus, during collision, collision force and fluid force have to balance each other. Due to the high accelerations during collision, the fluid force results for the most part from the acceleration of the fluid in the surrounding of the bubble, the added mass $m_{am} \approx 0.5\rho_f V_b$ [78]. For modelling the collision process, the added mass is assumed to be mounted to

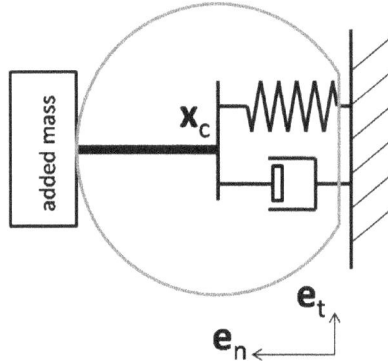

Figure 3.2: Equivalent network of the collision mechanics as assumed in this work. The added mass of the surrounding fluid is mounted directly to the centre position $\mathbf{x}_c$ of the bubble. Between bubble and obstacle elastic and diffusive forces appear.

the centre of mass of the bubble, as shown in the equivalent network in Figure 3.2. This results in the modified equation of motion

$$\mathbf{F}_{coll} = (F_{elastic} + F_{visc})\mathbf{e}_n + F_{tang}\mathbf{e}_t = m_{am}\ddot{\mathbf{x}}_c. \tag{3.6}$$

The collision force consists of three contributions: a normal elastic force $F_{elastic}$, a normal viscous force F_{visc} and a tangential viscous force F_{tang}. The latter is not sketched in Figure 3.2. The collision model has to provide values for F_{elastic}, F_{visc} and F_{tang} and add them in the equation of motion of the bubble (2.3). The right-hand side of Equation (3.6) is provided by the fluid solver, reacting on the movement of the bubble.

3.4.2 Elastic normal force

Similar to the spring model, a distance Δ is defined, as shown in Figure 3.3. It

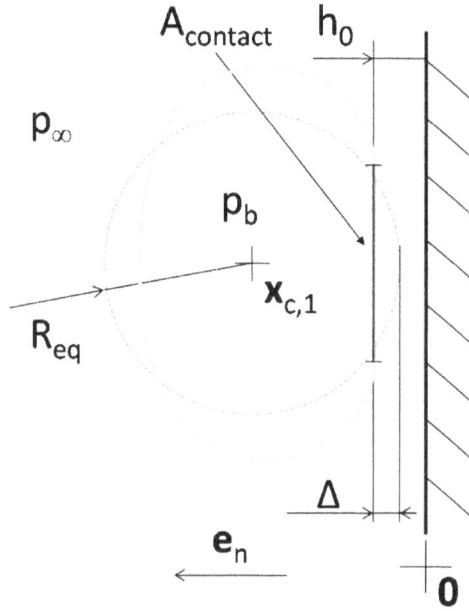

Figure 3.3: Relation between undeformed spherical bubble, as it is assumed by the numerical simulation, and deformed bubble in the collision model. The centre of mass $\mathbf{x}_{\mathrm{c}}$ and the bubble volume is congruent. The distance deficit Δ is the amount of deformation at the side of the obstacle.

results from comparing the deformed bubble with a sphere of the same volume, the same equivalent radius $R_{\mathrm{eq}}^3 = 3V_{\mathrm{b}}/4\pi$ and the same centre of mass. The amount by which the sphere is deformed by the wall represents Δ, the deficit in distance to the wall. It can be computed by $\Delta = R_{\mathrm{eq}} - \mathbf{x}_{\mathrm{c}} \cdot \mathbf{e}_{\mathrm{n}} + \Delta_0$

Elastic normal forces F_{elastic} result from variation of the surface energy E_{s} of the bubble due to deformation of the bubble when colliding with an obstacle. Thus, it can be derived by calculating the shape and shape deformation of a bubble during a collision process. The force then results from the total surface area A_{b}, the surface tension σ and the distance deficit Δ of the bubble

$$F_{\mathrm{elastic}} = \frac{dE_{\mathrm{s}}}{d\Delta} = \sigma \frac{dA_{\mathrm{b}}}{d\Delta}. \tag{3.7}$$

The shape of the bubble is computed by determining point-wise the principal curvatures R_α and R_β. They result from the Young-Laplace law comparing the external fluid pressure distribution p_{out} and the internal bubble pressure p_b and relating the difference to the mean curvature H_m

$$p_b - p_{\text{out}} = \sigma \left(\frac{1}{R_\alpha} + \frac{1}{R_\beta} \right) = \sigma H_m. \tag{3.8}$$

The pressure distribution p_{out} is not known, but assumed according to basic fluid mechanics. At this stage one may assume $p_{\text{out}} = p_s - \rho_f g y$, corresponding to the gravitational field only. The shape of a bubble with a certain volume V_b is to be computed. Figure 3.4 gives the flow chart for the iteration.

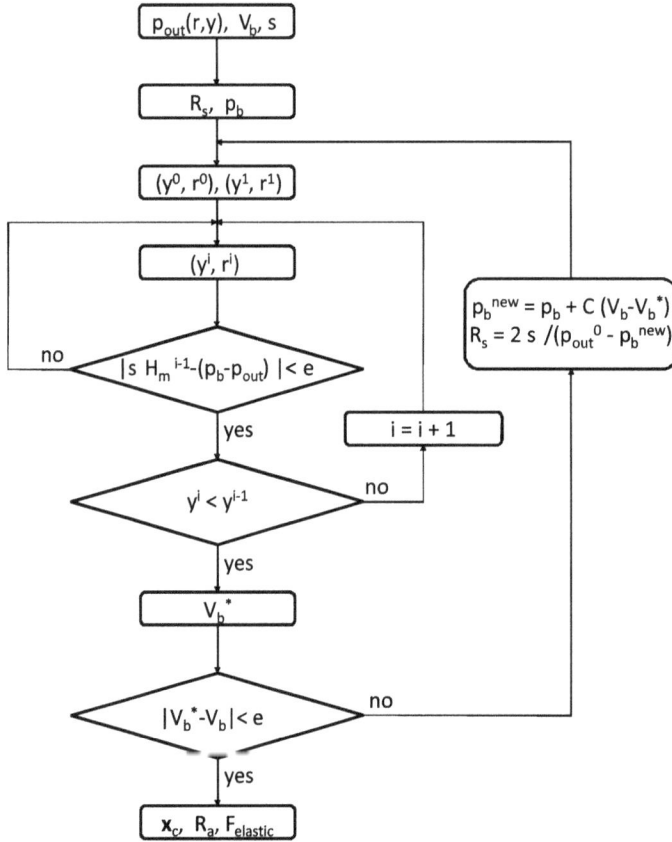

Figure 3.4: Flow chart of the procedure for determination of the shape of a bubble in contact to a horizontal wall.

From

$$R_s = (3V_b/4\pi)^{1/3}, \tag{3.9}$$

a starting guess R_s for the curvature radius at the apex is derived. From R_s a starting value for the bubble pressure $p_b = 2\sigma/R_s$ is derived. Also the step width for the point-wise computation $s_s = R_s/1000$ is chosen based on R_s.

Now, the computation starts at the lowest point of the bubble, marked in Figure 3.6. The coordinates of this point are assigned to be $(y^0, r^0) = (0, 0)$. The second point (y^1, r^1) results from the imposed curvature at the first point, R_s, the distance, s_s, and symmetry, as sketched in Figure 3.5

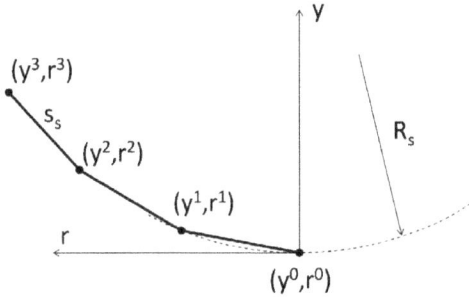

Figure 3.5: Sketch of the shape calculation. The calculation starts at the apex of the bubble with the guessed radius R_s. The distance between two discretisation points is s_s.

$$y^1 = s_s^2/(2R_s) \tag{3.10}$$
$$r^1 = s_s\sqrt{1 - y^1/(2R_s)}. \tag{3.11}$$

After knowing the first two points, each next point (y^i, r^i) can be calculated by a central difference approach with the help of the last two points (y^{i-1}, r^{i-1}) and (y^{i-2}, r^{i-2}). The mean curvature H_m^{i-1} can be derived from the three points $(i),(i-1)$, and $(i-2)$ assuming the surface to be a regular patch $\mathbf{x} : (r, y) \to \Re^3$ parametrisised by r and y

$$H_m = \frac{eG - 2fF + gE}{2(EG - F^2)}, \tag{3.12}$$

where E, F and G are the coefficients of the first fundamental form and e, f and g are coefficients of the second fundamental form of the parametrisation. Derivatives of the parametrisation are derived, using a central difference approach in the point (y^{i-1}, r^{i-1}). More information and the equations are given in [81]. The mean curvature has to satisfy the Young-Laplace law, given in Equation (3.8). Since Equation (3.12) is difficult to solve for (y^i, r^i), its identity is derived by an iterative procedure for each point. In that way, the shape of the bubble can be computed point-wise. When $y^i - y^{i-1}$ is smaller than 0 for the first time, the upmost point of the surface is found, corresponding to the contact point with the horizontal wall on top, as sketched in Figure 3.6. The corresponding radial position r^{i-1} yields the radius $R_a = r^{i-1}$ of the of the contact area. The surface of the bubble is assumed to continue with zero curvature, parallel to the horizontal wall and thus, forming a closed surface. According to the Young-Laplace law (3.8) zero curvature corresponds to zero pressure difference. This means, p_{out} at the contact area is not taken from the prescribed outer pressure distribution but is assumed to equal p_b. In reality, the

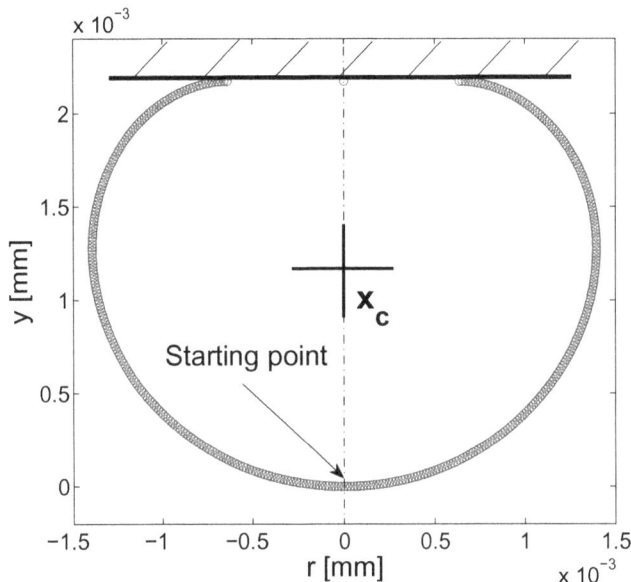

Figure 3.6: Numerical result of the shape of a bubble colliding with a horizontal wall.

pressure difference between p_{out} just outside of the gap and p_b inside the gap would either result in a flow from the gap, thinning the lamella or would be counteracted by disjoining pressure [1] between surfactants and the wall.

Now, that the surface shape is calculated, the volume of the resulting bubble is computed by

$$V_b = \pi/3 \sum_{i=2}^{max} (y^i - y^{i-1})(r^{i-1}r^{i-1} + r^i r^{i-1} + r^i r^i). \tag{3.13}$$

The resulting bubble volume is compared to the initially desired one, a new guessing value for p_b and R_s is derived and the construction of the bubble surface starts over again. In that way, by an outer iteration loop the desired bubble volume is achieved. Summing up, this algorithm allows to compute the equilibrium shape of a bubble of a certain volume V_b, exposed to a pressure field p_{out} and placed below a horizontal wall. Subsequently, the centre of mass, the surface area, the contact area and other geometrical measures can be derived from the shape. An example for such a shape is shown in Figure 3.6. It might be surprising, that this algorithm works so well. It is counter-intuitive, that the point where $y^i - y^{i-1}$ becomes negative actually corresponds to the physically correct contact point with the horizontal wall. In fact, in every loop of the outer iteration a physically correct shape is computed, because each computed shape fulfils the Young-Laplace law. The outer iteration loop only arranges the bubble volume to equal the prescribed one. The results can be validated after finding a shape, by analysing the equilibrium between contact force F_a and fluid force F_{out}

$$F_{elastic} = -F_{out} = -\int_{FluidArea} p_{out}\mathbf{e}_n dA \cdot \mathbf{e}_y = p_b \pi R_a^2 = F_a. \qquad (3.14)$$

This equilibrium is found to be valid with deviation below 1%, even though the contact force F_a depends quadratically on R_a and the uncertainty of R_a is a combination of the uncertainty of each surface point. The equilibrium of forces is a very convincing validation of the applied approach. In case of a purely gravitational pressure field $p_{out} = p_s - \rho_f g y$, also the buoyancy force $F_b = V_b \rho_f g$ can be considered for validation. It has to satisfy

$$F_b = V_b \rho_f g = \int_{FluidArea+ContactArea} p_{out}\mathbf{e}_n dA \cdot \mathbf{e}_y, \qquad (3.15)$$

which also was valid with a deviation below 0.1%. After deriving the shape of the bubble, we can derive the y-position of the centre of mass of the bubble $\mathbf{x}_c$. This yields the desired relation between the distance of $\mathbf{x}_c$ from the wall and the acting force $F_{elastic} = -F_{out} = F_a$.

The fluid force distribution p_{out}, acting on the bubble, of course depends on the dynamic collision process, and the instantaneous flow and pressure field around the bubble. With the above algorithm, however, an equilibrium state is computed. It is now investigated how much unknown fluctuations of p_{out} might influence the result. This is done byconsidering different functions $p_{out}(y, r)$.

$$p_{out} = p_s - \rho_f g y \qquad (3.16)$$

$$p_{out} = p_s + 0.5 \rho_f v_{ref}^2; \ \forall r^i > r^{i-1} \qquad (3.17)$$

$$p_{out} = p_s + \frac{3\mu_f v_{ref}}{D} \cos\theta; \ \theta = 2\mathrm{atan}(y/r); \ \forall r^i > r^{i-1} \qquad (3.18)$$

Equation (3.16) corresponds to a gravitational field, Equation (3.17) to a constant pressure on the lower hemisphere of the bubble, and Equation (3.18) to the solution of a potential flow with v_{ref} around the lower hemisphere. In case of Equation (3.17) and (3.18) on the upper hemisphere, p_{out} equals p_s.

From these tests, it is found, that the deformation of the bubble is, for physically reasonable distributions of p_{out}, only weakly dependant on the distribution but essentially only on the integral value F_{out}. Figure 3.7 shows the total fluid force F_{out} against the relative deficit of distance to the wall $\Delta/R_{eq} = (R_{eq} - \mathbf{x}_c \cdot \mathbf{e}_n)/R_{eq}$ for a deformed bubble, combined from several computations with different bubble parameters and force distributions $p_{out}(y, r)$. Obviously, different distributions of p_{out} collapse on the same graph. The one blue curve that does not fit in so well corresponds to Equation (3.17) with an artificially high velocity of $10 \times v_{ref}$. The other results are well represented by the fitting curve

$$\frac{F_{elastic}}{R_{eq}\sigma} = 18.5 \left(\frac{\Delta}{R_{eq}}\right)^2 + 2.0\frac{\Delta}{R_{eq}}. \qquad (3.19)$$

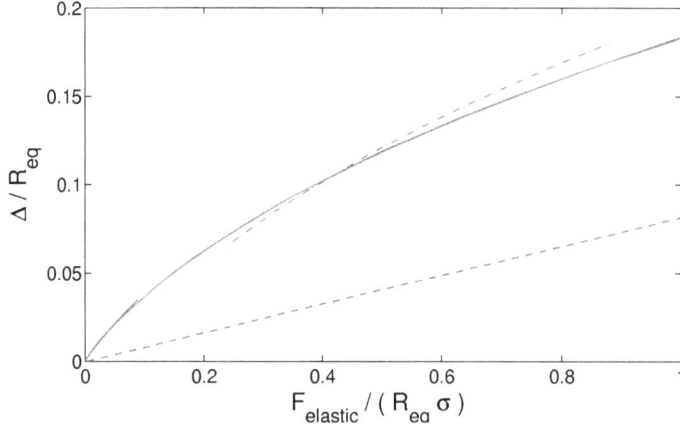

Figure 3.7: Relation between elastic collision force F_{elastic} and the distance deficit Δ between the centre of mass of a bubble and the obstacle for different force distributions (dash-dotted line). The solid line is the fitting curve according to Equation (3.19). The dashed curve shows the basic collision model from Equation (3.4).

This expression for the elastic force only involves the bubble position, its radius and the surface tension. It can be computed explicitly and is thus, easy to implement and requires low computational effort. The spring-like collision model, described in Section 3.2, is also shown in Figure 3.7 (broken line). It overestimates the stiffness of a bubble because it does not allow for relaxation from the prescribed collision shape in Figure 3.1

These elastic simulations also yield the relation between Δ and R_a, which are also nearly independent of the character of p_{out}. The curve can be fitted by

$$\frac{R_a}{R_{\text{eq}}} = 1.3 \frac{\Delta}{R_{\text{eq}}}^2 + 0.14 \frac{\Delta}{R_{\text{eq}}}. \tag{3.20}$$

This result will be used in Section 3.4.3.

3.4.3 Viscous normal force

When a particle moves normal to an obstacle with the velocity $u_c = -\dot{\mathbf{x}}_c \cdot \mathbf{e}_n$, additional forces act due to fluid interaction between particle and obstacle. Cox and Brenner [68] derived an expression for this lubrication force, solving the equations for Stokes flow between a plane wall and a spherical particle with distance δ between its surface and the wall

$$F_{\text{viscous}} = -6\pi\mu\dot{\mathbf{x}}_c \cdot \mathbf{e}_n R_{\text{eq}} \frac{R_{\text{eq}}}{\delta}. \tag{3.21}$$

However, this expression loses validity, when the particle is deformable and the lubrication force becomes very high. To estimate the limit of its validity, the pressure

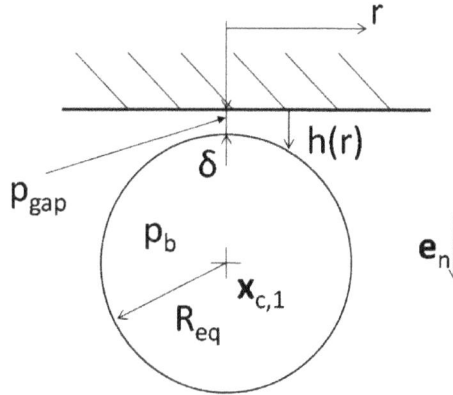

Figure 3.8: Sketch of the geometry of the gap when a spherical particle approaches a wall.

in the gap between sphere and wall can be used. Figure 3.8 sketches the geometry of the problem. The width of the gap denotes

$$h(r) = \delta + R_{eq} - \sqrt{R_{eq}^2 - r^2}. \tag{3.22}$$

The overpressure in the centre of the gap $p_{gap} - p_\infty$ between particle and wall can be calculated by integrating the pressure gradient from outside to the centre [68] yielding

$$p_{gap} - p_\infty = \frac{3}{2}\mu_f u_c \int_{R_{eq}}^{0} \frac{r}{h^3(r)} \, dr \approx \frac{3}{2}\mu_f u_c \frac{R_{eq}}{2\delta^2}. \tag{3.23}$$

When the pressure in the gap approaches the pressure inside the bubble, the surface between them decelerates. Consequently, the viscous force then becomes much smaller than Equation (3.21) suggests. The gap width h_0 at which gap pressure and pressure inside the bubble are equal, can be derived from Equation (3.23)

$$p_{gap} - p_\infty \approx \frac{3}{2}\mu_f u_c \frac{R_{eq}}{2h_0^2} = \frac{2\sigma}{R_{eq}}$$

$$h_0 = \sqrt{\frac{3}{8}\mu_f u_c \frac{R_{eq}^2}{\sigma}}. \tag{3.24}$$

To derive the viscous force, one needs to calculate the flow inside the gap. To do so, one also needs to calculate the shape of the lamella [77], which, again, results on the flow inside and outside the lamella. This ends up in a really complex problem. Here, instead, a simplification of this problem is proposed, sketched in Figure 3.9, which enables an analytic solution of the energy dissipation inside the collision zone.

The approach is as follows. A spherical bubble approaches a wall and maintains its spherical shape until the pressure in the gap equals the inner pressure of the bubble. Now, the gap width h_0 is frozen and deformation, according to Figure 3.6, takes

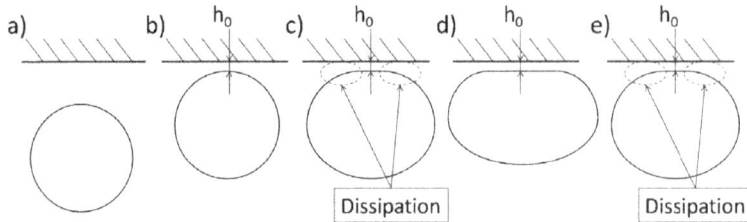

Figure 3.9: Proposed simplification of the collision process. (a) The bubble remains spherical while approaching an obstacle. (b) When the pressure in the gap equals the inner pressure of the bubble, the corresponding gap width h_0 is frozen. (c) Further approach of the bubble corresponds to deformation of the bubble, where the gap width remains constant while the gap area increases and liquid is displaced in the dissipation zone. (d) After reaching its minimum distance the process is reversed (e), sticking to the same geometry and deformation process.

place. After reaching the stagnation point, the bubble movement is reversed. The deformation of the bubble causes energy dissipation due to the viscous flow of the displaced liquid in the collision zone. However, since the gap width h_0 is constant, there is no dissipation in the lamella itself but only near its rim. To support this idea, the maximum energy dissipation $\Delta E_{\max}$ is estimated, that could take place inside the lamella if the whole lamella volume $V_{\max}$ is displaced with the maximum over pressure $p_{\max}$, yielding

$$\Delta E_{\max} = \Delta p_{\max} V_{\max} = \frac{2\sigma}{R_{\mathrm{eq}}} \pi \left(\frac{R_{\mathrm{eq}}}{2} \right)^2 h_0. \tag{3.25}$$

For a gas bubble in water with an assumed contact radius R_a of $R_a \approx R_{\mathrm{eq}}/2 = 0.5\,\mathrm{mm}$, $\Delta E_{\max}$ would equal about $2 \times 10^{-9}\,\mathrm{J}$. However, the typical kinetic energy $E_{\mathrm{kin,b}}$ of such a bubble and the surrounding fluid is

$$E_{\mathrm{kin}} = \frac{4}{3} \pi R_{\mathrm{eq}}^3 \rho_{\mathrm{f}} C_{\mathrm{am}} \frac{v_{\mathrm{ref}}^2}{2}. \tag{3.26}$$

With an added-mass coefficient C_{am} of 0.5 and a rise velocity v_{ref} of $0.234\,\mathrm{m/s}$ this would equal about $4 \times 10^{-8}\,\mathrm{J}$, i.e. 20 times bigger than the maximum dissipation in (3.25). Thus, thinning of the lamella does not seem to be the major dissipation mechanism.

In order to derive the energy dissipation in the collision zone, the geometry of the fluid displacement process is further simplified, as sketched in Figure 3.10. The lamella itself is supposed to be a plane disk of radius R_a and height h_0. Next to it appears the dissipation zone, that is bounded by a surface radius of R_{b}. Thus, one can easily derive the height of the gap in the collision zone as a function of the

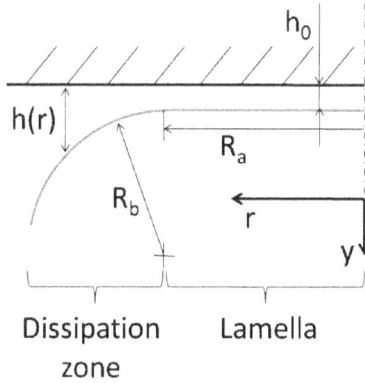

Figure 3.10: Proposed simplification of lamella geometry during collision. The thickness of the lamella, h_0, remains constant while its radius R_a changes, displacing liquid in the dissipation zone.

radial position r, yielding

$$h(r) \;=\; h_0 + R_b - \sqrt{R_b^2 - (r - R_a)^2} \tag{3.27}$$

$$\frac{dh}{dR_a} \;=\; \frac{-(r - R_a)}{\sqrt{R_b^2 - (r - R_a)^2}}. \tag{3.28}$$

When the bubble approaches the wall, R_a increases. The relation between R_a and Δ can be derived from the shape calculations in Section 3.4.2, yielding

$$\frac{\Delta}{R_{\text{eq}}} \;=\; 2.36 \left(\frac{R_a}{R_{\text{eq}}} \right)^2 + 0.04 \frac{R_a}{R_{\text{eq}}} \tag{3.29}$$

$$\frac{dR_a}{d\Delta} \;\approx\; 2.6 \frac{\Delta}{R_{\text{eq}}} + 0.14 \tag{3.30}$$

The radius R_b can also be derived from the shape calculator. It equals in good agreement the bubble radius R_{eq}.

The dissipative force F_{viscous}, acting on the bubble in the collision process, can be derived from the energy dissipation rate in the dissipation zone

$$F_{\text{viscous}} \;=\; \frac{dE}{d\Delta} = \frac{dE_{\text{dissipate}}}{dt} \frac{1}{u_{\text{c}}}. \tag{3.31}$$

The energy dissipation results from viscous flow of the displaced fluid in the gap. The volumetric flow through a vertical cross section at a given radial position $r = \tilde{r}$

results from the change in gap width h:

$$\frac{dV}{dt}\bigg|_{r=\tilde{r}} = u_c \frac{dR_a}{d\Delta}\frac{dV}{dR_a}\bigg|_{r=\tilde{r}} \tag{3.32}$$

$$\frac{dV}{dR_a}\bigg|_{r=\tilde{r}} = \int_{R_a}^{\tilde{r}} 2\pi r \frac{dh}{dR_a}\, dr \tag{3.33}$$

$$= 2\pi\left(R_a(h - h_0) + \frac{R_{eq}^2}{2}\mathrm{asin}\left(\frac{(\tilde{r} - R_a)}{R_{eq}}\right)\right.$$
$$\left. -\frac{\tilde{r} - R_a}{2}\sqrt{R_b^2 - (\tilde{r} - R_a)^2}\,\right) \tag{3.34}$$

Assuming a fully developed, two-dimensional Poiseuille flow in the gap, it is possible to derive the energy dissipation rate:

$$\frac{dE_{\mathrm{dissipate}}}{dt} = \int_{R_a}^{r_{\max}} \frac{dp}{d\tilde{r}}\frac{dV}{dt}\, d\tilde{r} \tag{3.35}$$

$$= \int_{R_a}^{r_{\max}} \frac{dp}{d\tilde{r}}\frac{dV}{dR_a}\, d\tilde{r}\,\frac{dR_a}{dt} \tag{3.36}$$

The pressure gradient in a Poiseuille flow is well known:

$$\frac{dp}{d\tilde{r}} = C_{\mathrm{bc}}\frac{12\eta_{\mathrm{f}}}{2\pi\tilde{r}h^3}\frac{dV}{dR_a}\frac{dR_a}{dt} \tag{3.37}$$

The parameter C_{bc} represents the boundary condition at the bubbles surface. For a no-slip boundary condition, $C_{\mathrm{bc}} = 1$, for free-slip, $C_{\mathrm{bc}} = 1/4$. Consequently, the energy dissipation can be written as:

$$\frac{dE_{\mathrm{dissipate}}}{dt} = C_{\mathrm{bc}}\frac{12\eta_{\mathrm{f}}}{2\pi}\left(\frac{dR_a}{dt}\right)^2 \int_{R_a}^{R_a+R_{eq}}\left(\frac{dV}{dR_a}\right)^2 \frac{1}{\tilde{r}h^3}\, d\tilde{r} \tag{3.38}$$

This integral can be solved analytically. However, the result is extremely complex and not at all handy. Therefore, by computing it numerically and fitting the result, the following numerical approximation was found

$$\int_{R_a}^{R_a+R_{eq}}\left(\frac{dV}{dR_a}\right)^2 \frac{1}{rh^3}\, d\tilde{r} \approx 4.0\sqrt{\frac{R_{eq}^3}{h_0}} + 3.0R_a\frac{R_{eq}}{h_0}. \tag{3.39}$$

For values $0.1\,\mathrm{mm} < R_{eq} < 0.3\,\mathrm{mm}$, $1\,\mu\mathrm{m} < h_0 < 100\,\mu\mathrm{m}$, $0 < R_a < R_{eq}$, Equation (3.39) approximates the integral with a deviation smaller than $10\,\%$. This deviation is negligible compared to the modeling uncertainties.

Combining Equation (3.31) with the Equations (3.38), (3.39), (3.29) and (3.30), one ends up with an expression for the viscous force, acting on a bubble in the collision process:

$$F_{\mathrm{viscous}} = u_c C_{\mathrm{bc}}\frac{12\mu_{\mathrm{f}}}{2\pi}(2.6\frac{\Delta}{R_{eq}} + 0.14)^2(4.0\sqrt{\frac{R_{eq}^3}{h_0}} + 3.0R_a\frac{R_{eq}}{h_0}) \tag{3.40}$$

This expression only relies on bubble position $\Delta = R_{eq} - x_c$, bubble velocity u_c, bubble radius R_{eq} and fluid viscosity μ_f. The lamella thickness h_0 can be estimated with Equation (3.24). The contact radius R_a is given in Equation (3.30) and only relies on R_{eq}, and Δ, as well. The force linearly depends on the bubble velocity, which is typical for viscous dissipation mechanisms. If the distance between bubble and wall is bigger than h_0, Equation (3.40) does not yield plausible results. Thus, the following scheme is applied(Table 3.1)

After the bubble centre reaches its closest position to the wall, its movement is

Table 3.1: *Scheme for switching between potential flow dissipation and the present method.*

Distance	Method
$\mathbf{x}_c \cdot \mathbf{e}_n - R_{eq} > h_0$	Equation (3.21)
$\mathbf{x}_c \cdot \mathbf{e}_n - R_{eq} < h_0$	Equation (3.40)

reversed. The dissipation process described above can be assumed to take place in reverse order. While the bubble withdraws from the obstacle, liquid is sucked in, an underpressure is generated in the dissipation zone. Due to this underpressure, the surface shape supposedly looks different than during the approach of the bubble, but this is not taken into account here. Instead, the geometry is assumed to be also described by Figure 3.10. Thus, the reversed Poiseuille flow dissipates energy according to the same mechanism, leading to exactly the same Equation (3.40) for the dissipative force F_{visc} as in the approaching process. The negative sign is provided now by the negative normal velocity u_c in the equation. Also, the process of detachment between bubble and obstacle is presumably different during the withdrawal of the bubble. But again, this is not taken into account here. When approaching, the bubble has a much bigger normal velocity, resulting in higher energy dissipation than during withdrawal of the bubble. Therefore, less exact modelling of the withdrawal process should not influence the overall process too much.

3.4.4 Tangential force

When a bubble moves parallel to a wall with the velocity $u_t = \dot{\mathbf{x}}_c \cdot \mathbf{e}_t$, there is an additional friction force, acting on the bubble. This situation is sketched in Figure 3.11. Goldman et al. [69] developed an equation for this force, assuming potential flow

$$F_{tang} = \frac{16}{5} \pi \mu_f u_t R_{eq} \ln\left(\frac{\delta}{R_{eq}}\right).$$ (3.41)

In the present collision model, the distance δ between bubble and wall is somewhat problematic, because the modelling of the normal collision force does not exclude a negative value of δ. Thus, using directly the distance δ can result in undefined values for the logarithm in Equation (3.41). To estimate a value for δ, Equation (3.24) could be used. However, if the bubble does not impact but glides parallel to the wall, the velocity of the bubble normal to the wall, that is required in Equation (3.24), is not

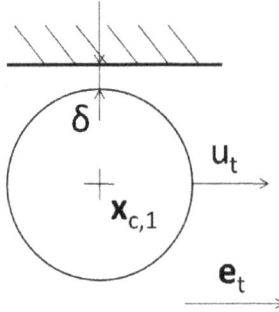

Figure 3.11: Geometry of a spherical bubble moving tangential to a wall or obstacle.

well defined. Thus, it is not recommended to use it. In this thesis, a limited value of $\mathbf{x}_c \cdot \mathbf{e}_n - R_{eq}$ is used:

$$\delta = \max(\mathbf{x}_c \cdot \mathbf{e}_n - R_{eq}, 10\,\mu\text{m}) \tag{3.42}$$

The value of $10\,\mu$m is chosen as a typical minimum lamella thickness, taken from [73]. If the bubble is far away from the wall, δ becomes larger than R_{eq} and Equation (3.41) yields positive values. This is physically incorrect and Equation (3.41) has to be switched off.

3.5 Testing and experimental validation

For comparison, the collision of a small bubble ($R_{eq} = 0.6\,\text{mm}$) with an inclined plate ($\alpha_w = 30°$) has been simulated with three different collision models. The setup is shown in Figure 3.12. The applied collision models are given in Table 3.3 and numerical parameters in Table 3.2

The resulting path, normal and tangential velocity are plotted in Figure 3.13.

Table 3.2: Parameters for the simulation of a bubble colliding with a tilted wall.

Quantity	Symbol	Value
Domain size	$L_x \times L_y \times L_z$	$6.4D \times 6.4D \times 6.4D$
Bubble diameter [m]	D	0.0012
Number of Bubbles [-]	N_b	1
Material properties	Water	Table 1.2
Spatial resolution [m]	Δs	6×10^{-5}
Time step [s]	Δt	10^{-4}
Lagrange points	n_L	1258
Tilt angle	α_w	$30°$

Obviously, the classical spring model Ia leads to oscillating movement. Using the elastic model (3.19) and the viscous dissipation term (3.40) results in a smoother

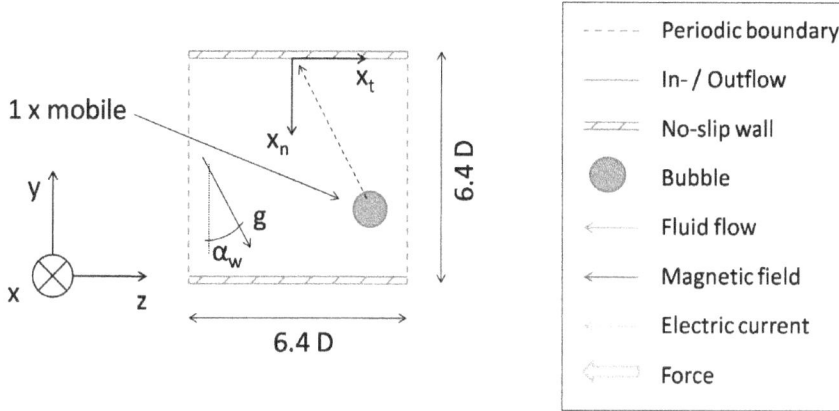

Figure 3.12: Numerical setup for the investigation of the collision of a bubble with a wall, tilted by α_w.

Table 3.3: Applied terms of the collision force for the comparison of three different collision models.

Case	F_{elastic}	F_{visc}	F_{tang}
Ia	$2\pi\sigma\Delta$	-	-
Ib	Equation (3.19)	Equation (3.40)	-
Ic	Equation (3.19)	Equation (3.40)	Equation (3.41)

path. Only one small rebound is visible. Including the tangential force term (3.41) does barely affect the normal position and velocity, but leads to a different tangential velocity after collision.

For validation, experimental results on a bouncing bubble are recomputed. In literature one can find some data for the trajectory of a bubble, colliding with a wall [82–84]. However, these experiments usually investigate large bubbles, undergoing strong deformations during the collision. Also, clean water is used most often, leading to a free-slip condition at the bubble surface. Thus, no literature data is found that could be used to validate the collision model derived above. Therefore, it was decided to design an own experiment, in order to have complete control of all parameters and to investigate configurations that correspond to the assumptions made above. In particular, $25\,\text{cm}^3$ soap were added to $5000\,\text{cm}^3$ water, leading to a highly contaminated bubble surface. Thus, the no-slip condition at its surface is valid [48]. The experimental parameters are given in Table 3.4. Figure 3.14 shows the experimental setup.

Small, monodisperse bubbles are created and transported into the fluid with a special microfluidic device [85]. They rise, reach terminal velocity and collide with the inclined plane. Before, they pass a light barrier that triggers the photo camera. Instead of taking a sequence of images, a stroboscope has been applied. It performs several flashes within one camera exposure time. Consequently, the same bubble appears several times on one picture. A sample picture is shown in Figure 3.15. For

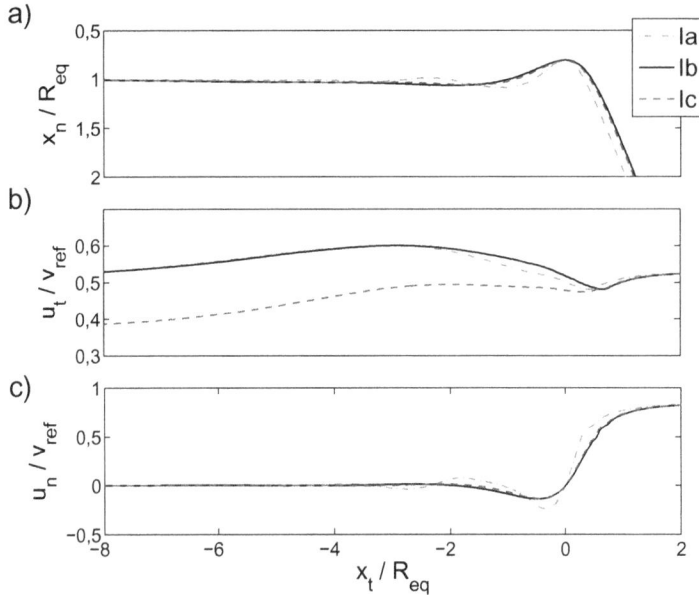

Figure 3.13: Normal centre position (a), tangential (b) and normal (c) velocity of a bubble, colliding with an inclined plane as a function of the tangential position. In time, the bubble follows the curves from right to left.

each setup, approximately 5000 pictures are taken. These images have been processed by applying edge filtering and a correlation-based detector for round objects, yielding the centre of mass for all bubble images within one picture. The velocity can be extracted from a central differences approach

$$u_i = \frac{x_{i+1} - x_{i-1}}{2 \, t_{\text{flash}}}. \tag{3.43}$$

The resulting data was filtered for plausible values with high Signal-to-Noise-Ratio. The inclination angle α_w of the plane and the bubble radius R_{eq} of the bubble has been varied as indicated in Table 3.4. At low α_w and low R_{eq}, the bubble smoothly propagates towards the plane without showing characteristic trajectories or rebounds. With increasing values of α_w or R_{eq}, a rebound after collision and a re-acceleration become more prominent in experiment and simulation, making the comparison more meaningful. However, at values of α_w above $45°$, the collision point varies from bubble to bubble, resulting in an unsharp measurement of the trajectory. At the same time, the ascension path of larger bubbles fluctuates, again resulting in an unsharp trajectory. Therefore, even though a large variety of parameters has been investigated, only the bubble diameter 0.6 mm is considered here and the inclination angle is limited to the small interval of $20° \le \alpha_w \le 45°$.

Figure 3.16 and 3.17 show experimental and numerical results of the path as well as, with respect to the plane, normal and tangential velocity of a bubble. In the experimental data of the velocities, Equation (3.43) causes a low-pass-filtering of the

Table 3.4: Geometry, material and optical parameters for the measurements of the collision process of a small bubble with an inclined plane.

Parameter	Value
Boxsize	$20 \times 20 \times 20\,\mathrm{cm}^3$
Bubble Diameter	$1.2\ldots4.4\,\mathrm{mm}$
Collision angle	$20\ldots45°$
Fluid density	$1000\,kg/m^3$
Material	Water (Table 1.2)
Optical resolution	$20\,\mu\mathrm{m}$
Flash rate	$250\,^1/_\mathrm{s}$
Flash length	$100\,\mu s$
Exposure time	$1/60\,\mathrm{s}$
Bubble rate	$1\,^1/_\mathrm{s}$

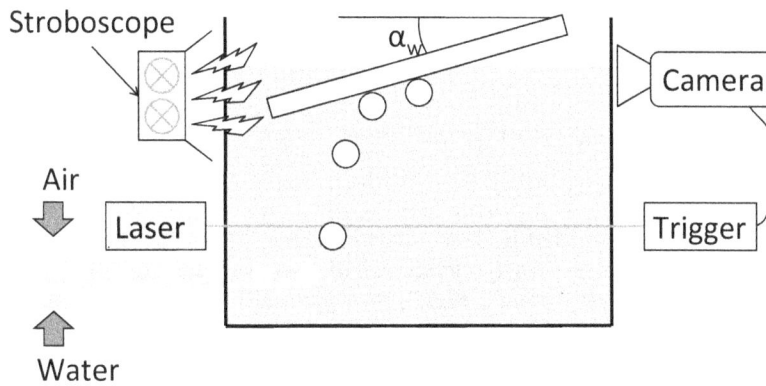

Figure 3.14: Experimental setup for measuring the trajectory of a bubble colliding with an inclined plane.

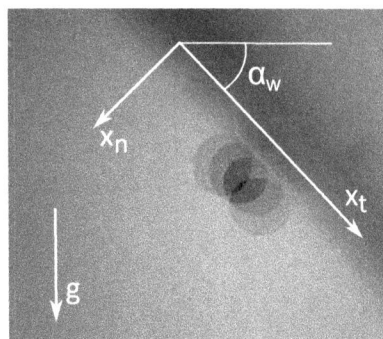

Figure 3.15: Sample image of a bubble ($R_{\mathrm{eq}} = 0.8\,\mathrm{mm}$) colliding with an inclined plane ($\alpha = 45°$).

velocities. In order to ensure better comparability, numerical data has been treated equally.

Experiment and simulations show fairly good agreement. The penetration depth,

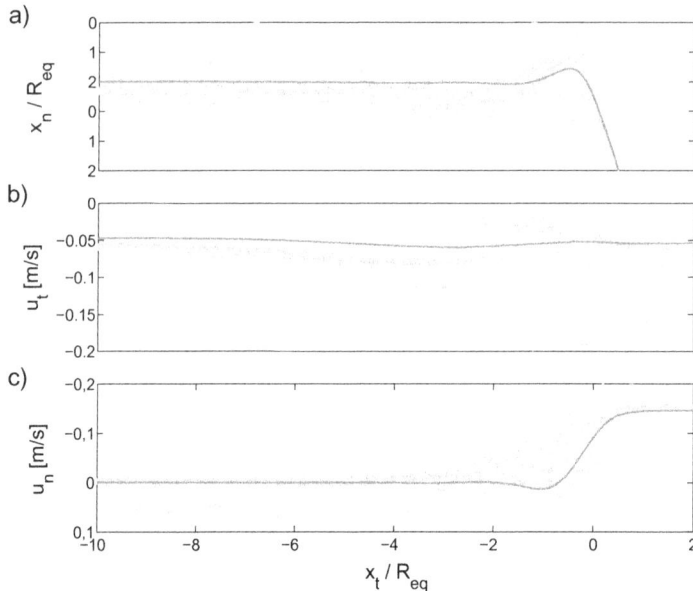

Figure 3.16: Experimental (points) and numerical (line) results of normal position (a), tangential velocity (b) and normal velocity (c) of a bubble with $R_{eq} = 0.8$ mm radius, colliding with a plane, inclined by $\alpha_w = 20°$, as a function of the tangential position.

the rebound height and the corresponding time-scales do roughly agree. The simulation reproduces the fact, that there is only a short contact time and nearly no rebound at low angle, while at higher angle longer contact times and a rebound take place. Also the tangential velocity shows good agreement. In simulation as well as in experiment, one can see the deceleration in the moment of impact, the partial acceleration during the rebound and then a slow declaration towards the terminal tangential velocity. Oscillations, as predicted by the spring model in Figure 3.13 do not take place.

Of course, the agreement between experiment and simulation is not perfect, but one has to keep in mind that the advanced collision model does not include any empirical parameter that potentially could be used to calibrate the collision model for better agreement with the experiment.

3.6 Conclusions

In this chapter, a basic and an advanced model for the collision force, acting between a bubble and a wall or another bubble, have been derived. To apply one of these models, one does not need to know anything about pressure or velocity of the fluid

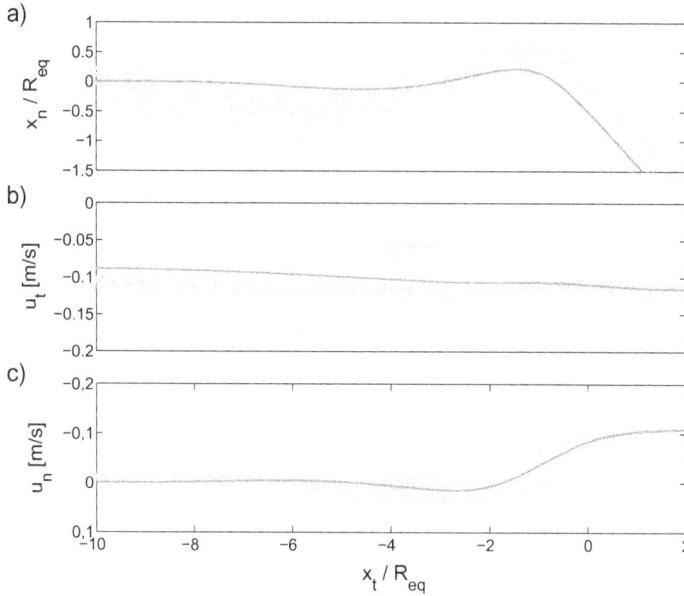

Figure 3.17: Experimental (points) and numerical (line) results of normal position (a), tangential velocity (b) and normal velocity (c) of a bubble with $R_{eq} = 0.8\,\text{mm}$ radius, colliding with a plane, inclined by $\alpha_w = 45°$, as a function of the tangential position.

in the gap between bubble and obstacle. Instead, the models only employ position and velocity of the bubble. Both models also exclude empirical parameters that would have to be calibrated in order to represent reality. Instead, parameters are derived from material properties and physical approximations.

Comparing the basic and the advanced collision model reveals an overestimation of the bubble stiffness by the basic model. Due to the absence of dissipative forces, the basic model also results in stronger bubble oscillations after an impact, that are not reproduced by the validation experiment. These oscillations might lead to artificial movements of bubbles within a cluster. Therefore, in case of strong compression of many bubbles, the advanced collision model should be used, if the basic model yields oscillatory movement of bubbles.

4 Rise and agglomeration of bubbles

4.1 Motivation

In the first step, the unmodified rise and agglomeration of spherical bubbles is investigated. The rise or sedimentation of single particles and highly disperse particle clusters is well investigated [86–89]. Also the drainage through clusters of densely packed particles is well understood [90]. In this chapter, both regimes as well as the transition between them are investigated. The aim of this chapter is to validate the approach, to get familiar with the abilities and restrictions of the numerical procedure and to derive typical scales.

4.2 Rise of a single bubble

In order to set up proper simulations, one needs to know certain information about the rise of a bubble for the considered set of parameters. From the acceleration of the bubble one can derive the necessary height of the domain to ensure that rising bubbles reach terminal velocity v_{ref} before the impact on the foam cluster. The height and the rise velocity determine the physical laps of time one has to simulate until all bubbles reach the cluster. Therefore, the rise of an initially steady bubble is simulated. The setup is shown in Figure 4.1. The corresponding parameters are given in Table 4.1. The resulting trajectory of the bubble is shown in Figure 4.2.

Table 4.1: Parameters for the simulation of static drainage.

Quantity	Symbol	Value
Domain size [m]	$L_x \times L_y \times L_z$	$0.016 \times 0.032 \times 0.016$
Bubble diameter [m]	D	0.002
Number of Bubbles [-]	N_{b}	1
Material properties	Aluminium	Table 1.1
Spatial resolution [m]	Δs	1.25×10^{-4}
Time step [s]	Δt	10^{-4}
Lagrange points	n_{L}	806

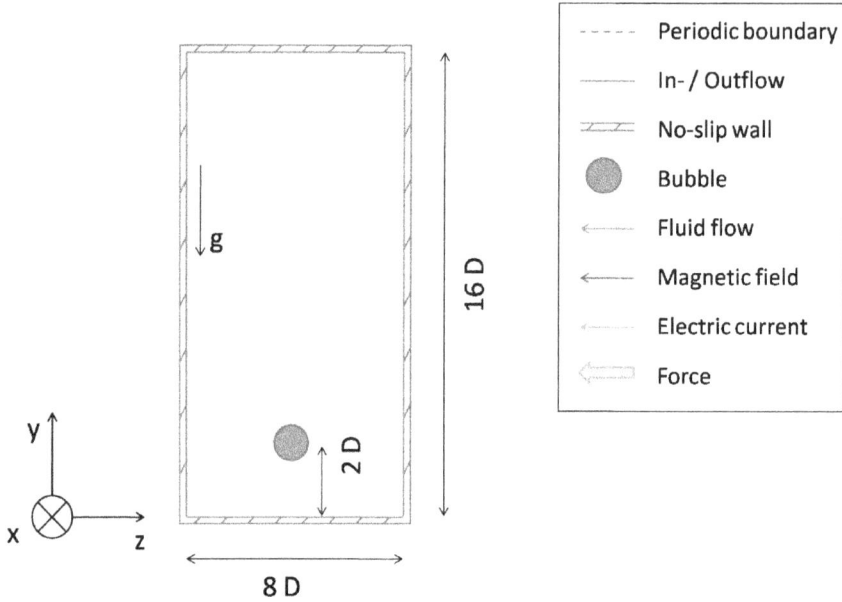

Figure 4.1: Numerical setup for the simulation of a single, rising bubble.

From this trajectory several information can be derived. The terminal velocity of the bubble, which will be used as reference velocity and therefore labelled v_{ref} equals $v_{\text{ref}} = 0.234\,\text{m/s}$. After $0.06\,\text{s}$ the bubble reaches terminal velocity. But, more important, after a rising distance of about $4\dots6D$ the bubble reaches terminal velocity. Setups that investigate the agglomeration of freely rising bubbles should therefore also include this length of free rising path for the bubbles in order to get the rising velocity and therefore, the impact physically correct. In case of limited computational power, $2D$ of rising length might be acceptable.

The terminal velocity v_{ref} corresponds to a bubble Reynolds number of $Re = v_{\text{ref}}D\nu_{\text{f}} = 936$. The Eötvös number of a gas bubble of $2\,\text{mm}$ diameter in liquid aluminium equals $Eo = g(\rho_{\text{f}} - \rho_{\text{b}})D^2/\sigma = 0.094$. With both numbers one can localise the shape of the rising bubble in the regime map of Clift et al. [48], shown in Figure 4.3. The bubble is situated in the spherical regime. Hence, it is supposed to maintain its spherical shape during ascension. For a sphere moving through quiescent fluid the drag coefficient c_{D} in dependency of the Reynolds number is well known [48]. For a sphere with no-slip condition on its surface, moving at a Reynolds number of $Re = 936$ the drag coefficient is supposed to equal [48]

$$\log_{10}c_{\text{D}} = 1.6435 - 1.1242\log_{10}Re + 0.1558\left(\log_{10}Re\right)^2 = 0.4772. \qquad (4.1)$$

It can also be derived from the simulation, assuming equilibrium between drag force

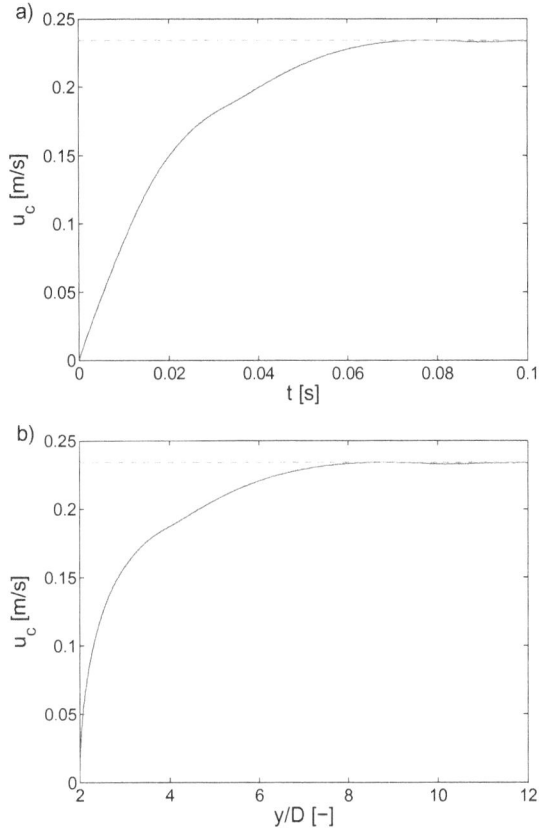

Figure 4.2: Trajectory of a rising bubble. Rising velocity over time (a) and rising velocity over vertical position (b). The broken line marks the terminal velocity v_{ref}.

F_{D} and buoyancy force F_{b}, i.e.

$$
\begin{aligned}
F_{\mathrm{b}} &= \frac{\pi}{6} D^3 (\rho_{\mathrm{f}} - \rho_{\mathrm{b}}) g = F_{\mathrm{D}} = c_{\mathrm{D}} \frac{\pi}{4} D^2 \frac{\rho_{\mathrm{f}}}{2} v_{\mathrm{ref}}^2 \\
c_{\mathrm{D}} &= 0.4777
\end{aligned}
\tag{4.2}
$$

This appears like a perfect match between simulated drag coefficient and literature. The literature value, however, is derived by fitting several sets of data, showing about $5\ldots 10\%$ variation. The close agreement observed here should therefore not be over-emphasized. In any case, it demonstrates the validity of the simulation. The numerical resolution in this investigation is relatively low with $D/\Delta s = 16$ grid points per diameter. Throughout this thesis, this is the minimum resolution that is used. This is also in line with the experience from [62] and [52]. Often, resolutions of 20 or even 40 grid points per diameter are applied throughout this thesis, in order to ensure resolution of small channels between aglomerated bubbles.

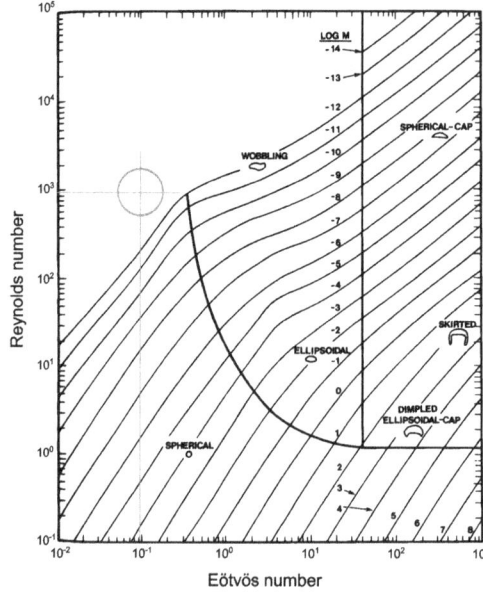

Figure 4.3: Shape regimes for a rising bubble [48]. The marked point corresponds to a single gas bubble with 2 mm diameter rising in liquid aluminium.

4.3 Static drainage

After investigating the rise of a single bubble, now the interaction of many bubbles is simulated. In order to investigate the transition of a cluster of bubbles from the floating or rising regime to the agglomerated regime in some kind of steady state setup, the static drainage experiment of [91] is devised here. A fixed number of bubbles is situated in a vertical column. On top, a certain amount of fluid per second $\dot{V}_0$ is added. The fluid drains downward through the foam, resulting in a certain, constant gas fraction Φ of the foam. In this work, the approach is extended to very high volumetric flows $\dot{V}_0$, causing the bubbles even to float.

The setup of these simulations is shown in figure 4.4. Parameters are given in Table 4.2. Homogeneously distributed, 250 spherical bubbles are placed in a domain with periodic boundaries in both horizontal directions. A constant, vertical drainage velocity $v_0 = \dot{V}_0/(L_x L_z)$ is imposed at the upper boundary of the domain, channelling downward to the outlet. Near the top, an artificial boundary for the bubbles is introduced in order to prevent the bubbles from leaving the domain. The pressure drop Δp between top and bottom of the domain, averaged over the horizontal cross section is recorded, with the hydrostatic pressure removed. The gas fraction is determined in an average sense from

$$\Phi = \frac{n_{\mathrm{b}} V_{\mathrm{b}}}{(y_{\max} - y_{\min}) L_x L_z}, \tag{4.3}$$

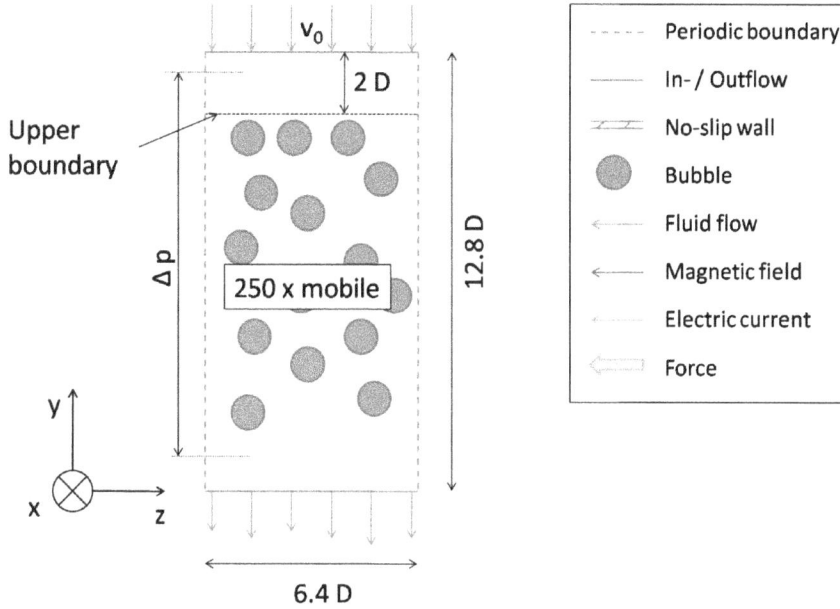

Figure 4.4: Setup of the static drainage simulation.

where y_{max} is the average y-position of the centres of the five highest bubbles and y_{min} of the five lowest ones.

Figure 4.5 shows the pressure drop and the gas fraction as a function of the static drainage velocity. Basically, two regimes exist. For drainage velocities below $v_0 = 0.03 \, \text{m/s}$, the pressure drop linearly increases with the drainage velocity while the gas fraction remains more or less constant. For drainage velocities above $v_0 = 0.03 \, \text{m/s}$, the pressure drop remains constant while the gas fraction increases non-linearly with the drainage velocity. Figure 4.6 shows a snapshot of the flow in both regimes. Left, at low drainage velocity, the bubbles agglomerate and the fluid flows downward through small channels. Right, at high drainage velocities, the bubbles are fluidised, occupying a larger volume and allowing the fluid to pass in wider channels.

4.4   Agglomerated regime

The agglomerated regime corresponds to drainage through sediment. Darcy's law [90] describes the relation between pressure drop and drainage velocity

$$\frac{\dot{V}}{L_x L_z} = v_0 = k_f \frac{\Delta p}{\rho_f g H_{\text{packing}}}, \qquad (4.4)$$

Table 4.2: Parameters for the simulation of static drainage.

Quantity	Symbol	Value
Domain size [m]	$L_x \times L_y \times L_z$	$0.0128 \times 0.0256 \times 0.0128$
Bubble diameter [m]	D	0.002
Number of Bubbles [-]	N_b	250
Material properties	Aluminium	Table 1.1
Spatial resolution [m]	Δs	10^{-4}
Time step [s]	Δt	10^{-4}
Lagrange points	n_L	1256
Drainage velocity [m/s]	v_0	$0 \ldots 0.05$

with k_f the hydraulic permeability and H_{packing} the size of the sphere cluster in y-direction. The value for k_f can be estimated with the mean sphere diameter D [92]

$$k_f \approx 0.0116 \left(\frac{D}{1\,\text{mm}}\right)^2 \frac{\nu_{\text{water}}}{\nu_f} 1\,\text{m/s} = 0.093\,\text{m/s}. \tag{4.5}$$

The thickness of the packing H_{packing} is derived, similar to the gas fraction from $H_{\text{packing}} = y_{\text{max}} - y_{\text{min}}$ for the case of $v_0 = 0\,\text{m/s}$, yielding $H_{\text{packing}} \approx 6.9D$. With these values, Equation (4.4) yields a ratio between pressure drop $\Delta p / \rho_f$ and drainage velocity v_0

$$v_0 = 0.093\,\text{m/s} \frac{\Delta p}{\rho_f g 5.4 D} \approx \frac{\Delta p}{\rho_f} 0.68\,\text{s/m}. \tag{4.6}$$

The corresponding graph is plotted in Figure 4.5 a) with a green line. The data from the simulation gives higher values for the pressure drop, which might result from the fact, that the flow is not completely viscous because the Reynolds number $Re = Dv_0/\nu_f$ for $v_0 = 0.03\,\text{m/s}$ equals 120 which means that inertia effects are not completely negligible.

4.5 Floating regime

For drainage velocities above $v_0 = 0.03\,\text{m/s}$, the pressure drop remains more or less constant at $\Delta p = 0.07\,\text{m}^2/\text{s}^2$. This value is in good agreement with the buoyancy force of all bubbles, distributed on the horizontal cross section of the domain:

$$\frac{\Delta p}{\rho_f} \approx \frac{(\rho_f - \rho_b)}{\rho_f} \frac{n_b V_b g}{L_x L_z} \approx \frac{(\rho_f - \rho_b)}{\rho_f} 0.064\,\text{m}^2/\text{s}^2 \approx 0.07\,\text{m}^2/\text{s}^2 \tag{4.7}$$

This means, the bubbles are floating. The hydrodynamic force on a bubble equals its buoyancy force. Therefore, each bubble causing a pressure drop that refers to its buoyancy force. This regime can be described by the conservation of momentum in y-direction

$$\int_\Omega \frac{\partial \rho \mathbf{u}}{\partial t} + \oint_{\delta\Omega} \rho \mathbf{u}(\mathbf{u} \cdot \mathbf{e}_n) dA = -\oint_{\delta\Omega} p \mathbf{e}_n dA + \oint_{\delta\Omega} \hat{\tau} \mathbf{e}_n dA + \int_\Omega \mathbf{f} dV \tag{4.8}$$

$$0 + \rho_f v_0^2 L_x L_z - \rho_f v_0^2 L_x L_z = -\Delta p L_x L_z + 0 + n_b F_b \tag{4.9}$$

a)

b)

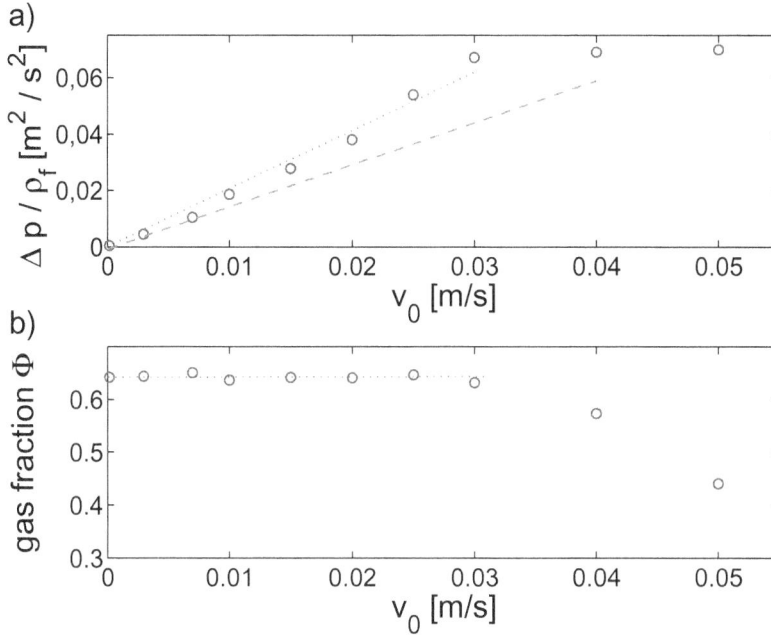

Figure 4.5: Pressure drop (a) and gas fraction (b) as a function of the static drainage velocity. The blue dotted lines represent linear fits to the data with drainage velocities below 0.03 m/s. The green broken line visualises Darcy's law.

The first term on the left-hand side of Equation (4.8) is zero due to steady conditions. The second term vanishes because the outflow is chosen at a distance far enough from the bubbles to allow the flow to equalise. Due to conservation of mass the average outflow velocity has to equal the inflow velocity v_0. Flows across the vertical boundaries automatically balance their momentum because of the periodic conditions. The first term on the right-hand side of Equation (4.8) represents the pressure drop across the domain. The second term is zero because, again, at the outflow the flow has equalised, resulting in low shear forces. Across the periodic boundaries shear forces balance as well. The third term represents the sum of the buoyancy forces of all bubbles. Summing up, a balance between pressure drop and buoyancy exists. An increase of the drainage velocity in this regime will not increase the pressure drop but results in a lower gas fraction, because the bubbles give way to the fluid flow until the flow channels are wider and the balance of forces is maintained.

Figure 4.6 shows a snapshot of the flow in both regimes. Left, at low drainage velocity, the bubbles agglomerate and the fluid flows downward through small channels. Right, at high drainage velocities, the bubbles are fluidised, claiming a higher volume and allow the fluid to pass in wider channels.

v₀ = 1 cm/s   v₀ = 5 cm/s

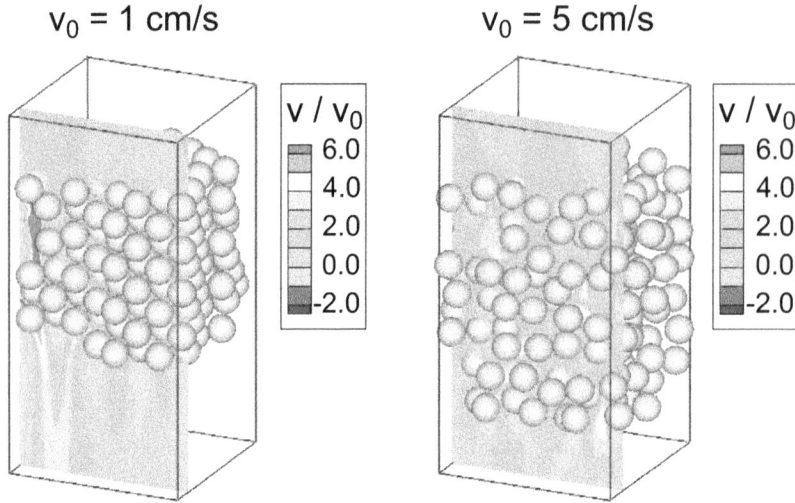

Figure 4.6: Snapshot of the bubble positions and the corresponding downward fluid velocity component for low (left) and high (right) drainage velocity.

4.6   Structure formation

The author, Anniina Salonen and colleagues [93] investigated experimentally the sedimentation of small sediment particles in tiny vertical capillaries. Sedimentation is an upside-down situation of the floating regime in static drainage. In both cases the particles experience a relative flow velocity. In both cases the relative flow causes drag forces that compensates the particles buoyancy or gravitational force, resulting in vanishing acceleration of the particle. The differences between the floating regime and sedimentation is the high density ratio of sediment compared to bubbles, and the velocity boundary condition at the vertical wall. Figure 4.7 illustrates both cases.

 In sedimentation it is found, that the sediment does not remain equally distributed throughout the cross-section of the vertical capillary but concentrates in one part of the cross section. In that way, it moves downward in a lane, avoiding the displaced liquid that flows upward in the opposite lane. Consequently, the sedimentation speeds up. Under certain conditions, the sedimentation velocity even exceeds the Stokes velocity of a single sediment particle. Figure 4.8 shows the distribution and downward velocity of sedimenting plastic beads in glycerol, visualising the formation of lanes. Further analysis of this feature reveals, that confinement in both horizontal direction is necessary. Also, the Reynolds number of the particle has to be sufficiently low. In the static drainage setup both requirements are not fulfilled. The horizontal boundary conditions are periodic, corresponding to the absence of confining walls. The bubble Reynolds number equals about 1000, which is far above the Reynolds number of the sedimentation investigated, that equals about 0.02. For comparison

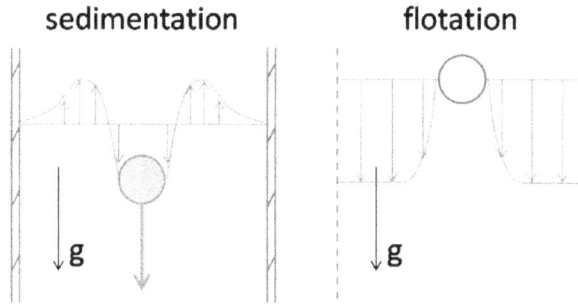

Figure 4.7: Sketch of the principal velocity distribution in sedimentation (left) and flotation of bubbles (right).

also the sedimenting process of small plastic beads in water is investigated, yielding a Reynolds number of 900. In this case no formation of lanes and no speed-up is found.

In the simulation of the floating state, 250 bubbles move relative to the fluid. This raises the question, whether the bubbles concentrate in a part of the cross section of the channel to avoid the moving fluid. Figure 4.9 shows a snapshot of the distribution of gas fraction, averaged over the y-direction. Obviously, the distribution is not perfectly homogeneous but shows local fluctuations. However, the bubbles do not concentrate in a part of the cross section of the channel, as observed in the experiments at low Reynolds number. This is in line with the experimental findings, because at high Reynolds numbers in the experiment no formation of lanes was found neither.

4.7   Conclusions

In the static drainage simulations two regimes exist. Below a drainage velocity of $0.03\,\mathrm{m/s}$, for the present set of parameters the bubbles agglomerate in the upper region of the domain. Above this drainage velocity, the bubbles become mobilised and float. The simulated drainage resistance in the agglomerated regime is in good agreement with literature, showing the applicability of the method. This is important, because in the agglomerated regime, the interstitial channels become relatively small, as shown in Figure 4.10

This leads to a small number of numerical grid points across the channels and therefore, the channel flow might not be resolved well. However, the channel flow is laminar. The bulk Reynolds number in such a channel at $v_0 = 0.03\,\mathrm{m/s}$ and

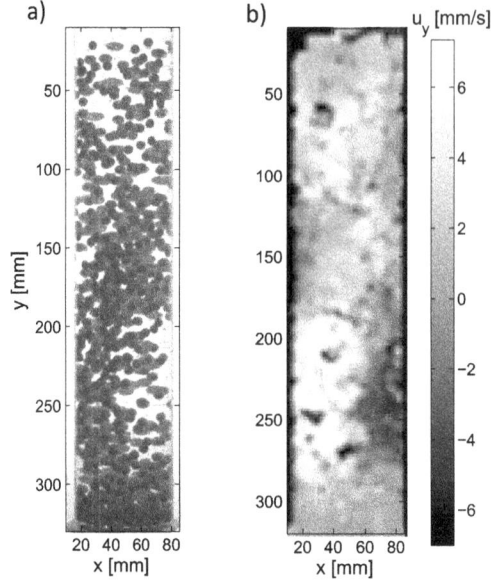

Figure 4.8: Distribution (a) and downward velocity (b) of plastic beads sedimenting in glycerol. The formation of lanes is visible, speeding-up the sedimentation process.

$D = 2\,\mathrm{mm}$ equals about

$$
\begin{aligned}
Re &= uL\frac{1}{\nu_\mathrm{f}} \\
A_2 &= \frac{\sqrt{3}D^2}{4} \\
A_1 &= A_2 - \frac{\pi D^2}{8} \\
u &= v_0\frac{A_2}{A_1} = 0.322\,\mathrm{m/s} \\
L &= \frac{4A_1}{0.5\pi D} = 2.05 \times 10^{-4}\,\mathrm{m} \\
Re &= 132.
\end{aligned}
\tag{4.10}
$$

Therefore, the small amount of grid points is sufficient to yield the correct pressure drop.

In the floating regime the pressure drop is in balance with the collective buoyancy force of all bubbles. An increase in drainage velocity then does not result in a higher pressure drop, but the bubbles give way to the fluid by increasing their distance an occupying more space, which results in a decrease of the gas fraction. In sedimentation under confinement and at very low Reynolds numbers a formation of lanes is found, increasing the sedimentation speed significantly. However, in the simulations of floating bubbles no such lane formation is found. The bubbles distribute equally in the x-z-plane.

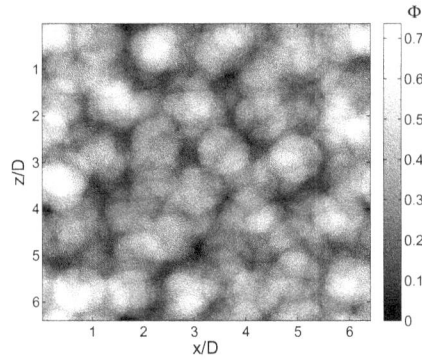

Figure 4.9: Snapshot of the gas fraction (vertically averaged) of 250 bubbles in the floating regime. No formation of lanes is visible.

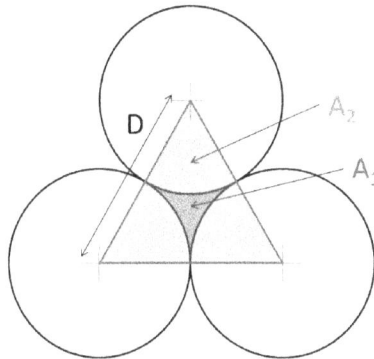

Figure 4.10: Geometry of a small channel between three spheres.

In the agglomerated regime, the formation of crystalline arrangement of the bubbles is observed. For some reason, usually face-centred cubic arrangement appears. In the next chapter, this preference will be further investigated.

5 Bubble crystals

5.1 Motivation

This chapter bases on the author's publication [94] and deals with the formation of crystalline structures in the agglomeration of spherical bubbles. In Figure 4.6, left, one can see that the bubbles form a regular, crystalline arrangement. Of course, this is only possible, because the bubbles are monodisperse, which means they all have the same diameter. However, these structures are not a special feature of this idealized computational simulation but also have been observed in many experiments with spherical monodisperse objects of different kind [3, 28, 95–101]. Under the influence of gravity these objects try to pack as densely as possible, forming close-packed structures. Basically there are two crystalline packing structures that both offer the most dense sphere packing of $\Phi = 0.74$. These are face-centred cubic (FCC) and hexagonally close-packed (HCP). As pictured in Figure 5.1 both structures are

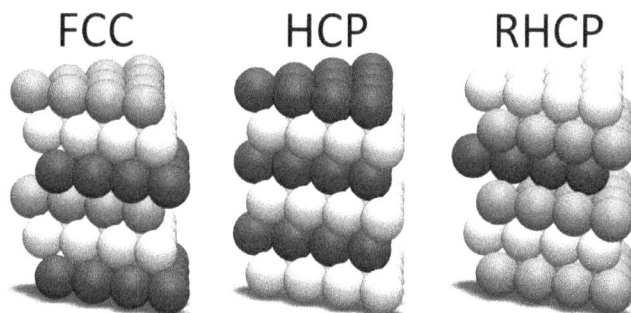

Figure 5.1: Structure of face-centred cubic (FCC), hexagonally close-packed (HCP) and random close-packed (RHCP) sphere packing.

very similar. They both consist of layers of triangularly close-packed spheres. Only the stacking sequence is different. In FCC the layers are arranged in an -A-B-C-A-B-C-A- sequence, in HCP the sequence is -A-B-A-B-A-. Random hexagonally close-packed refers to a random stacking sequence, as pictured in Figure 5.1 right. But despite this sequence difference, FCC and HCP have many properties in common, e.g. the packing density and the number of direct neighbours for each bubble. Nevertheless, several researchers found a preference of FCC over HCP packing when packing spherical objects such as glass beads [101], Microbubbles [3, 28, 95–97], colloids [98, 99] and nanoparticles [100].

In a microbubble system v.d.Net et al. [28, 95, 96] spent considerable effort to quantify the preference of FCC, resulting in a ratio of roughly 65 to 35. However, the reason for the preference of FCC was not discovered. In colloids, the slightly higher entropy of FCC packing is often used as an argument [102]. However, for macroscopic bubbles or even heavy glass beads this is very unlikely. Another hypothesis takes into account the difference in the distance between spheres that are not direct neighbours. In HCP packing the smallest distance between one sphere and a sphere from the second next layer equals $1.63\,D$ while in FCC it equals $1.73\,D$. If spheres are repelling each other, they would therefore prefer FCC packing. However, in case of bubbles there is no long-range interaction between them as is the case, e.g. for charged colloids [98]. Therefore, the distance to the second-closest neighbours should not play any role. In bubble packing, there is the hypothesis, that hydrodynamic interactions between bubble and existing cluster in the agglomeration process might cause the FCC preference. The mechanism is sketched in Figure 5.2. When

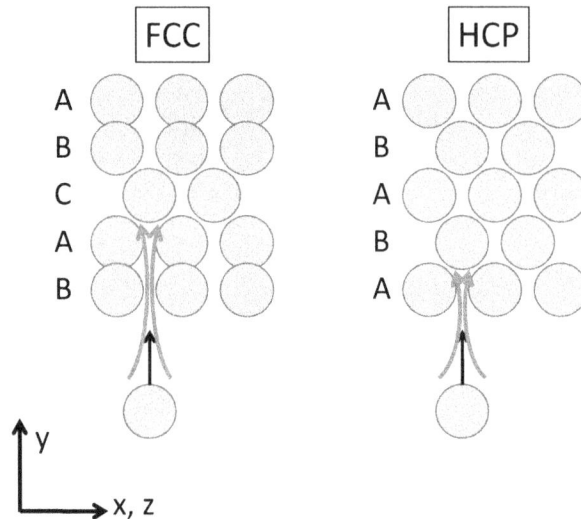

Figure 5.2: Two dimensional sketch of the hydrodynamic mechanism that is suspected to oauoo FCC proforonoo. Addod mass of the rising bubble can penetrate more difficult in HCP arrangement, resulting in the hydrodynamic distortion into FCC arrangement.

a bubble rises, it transports some liquid in its surrounding. This liquid, the added mass, dominates the inertia of the light bubble. When the bubble approaches an already existing cluster, it might prefer an FCC position, because in this position there is a gap in the next two layers above, which allows the added liquid to diffuse more easily into the cluster. On the contrary, in HCP this gap is blocked in the second-next layer by another bubble. Therefore, the added liquid cannot diffuse into the cluster with the same ease as for FCC and a region of higher pressure is generated above the bubble by the impact of the added fluid on the cluster. This

higher pressure might repel the bubble sidewards into an FCC position. The validity
of this hypothesis will be investigated below.

5.2   Setup

In this work, the preference of FCC packing has been reproduced numerically. The
simulations, described in Chapter 4 above, usually form crystalline packings, showing a statistical preference of FCC. This is remarkable, because it means that the
physical effect that causes the preference is also covered by this type of simulation.
This excludes, for example, electrostatic or other long-range forces, because these
are not considered in the simulation.

After finding the preference, a special setup is designed in order to enhance crystal
growth. To do so, layers of hexagonally close-packed, fixed bubbles are placed in
the upper part of the domain, triggering growth of crystalline packing. To avoid
disturbing influence of the side walls, periodic boundary conditions are used in both
horizontal directions. This requires that the horizontal dimensions L_x and L_z of the
domain correspond to the periodic length of the crystal that is to be bred. At the
same time, PRIME requires that periodic directions are discretized with a number
of grid points that is a multiple of 2 for technical reasons. Since the grid is isometric,
the ratio of the size of the domain in periodic directions x and z is also restricted.
It has to satisfy $L_x/L_z = 2^a$ with a an integer. It is a lucky coincidence that a
nearly perfect solution for this problem exists, that does not require extremely large
domains. It is sketched in Figure 5.3.

The domain length in $x-$direction equals $L_x = 7(D + \Delta_0)$, which is a multiple of

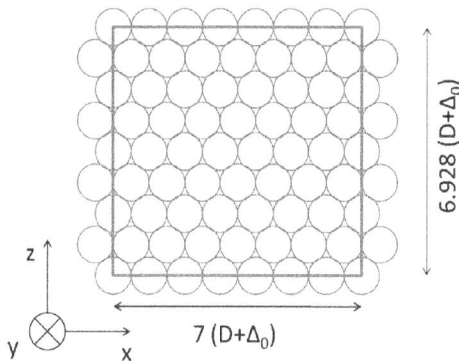

*Figure 5.3: Sketch of the arrangement of a horizontal layer in close-packed ordering,
neglegting the distance offset Δ_0.*

the periodic length in x-direction $(D+\Delta_0)$. The domain length in $z-$direction is set
to equal $L_z = 7(D + \Delta_0)$ as well, in order to satisfy numerical restrictions. At the
same time, the periodic length in $z-$direction equals $\sqrt{3}(D + \Delta_0)$ and 4 times this
periodic length equals $6.928(D + \Delta_0)$ which nearly perfect fits to $L_z = 7(D + \Delta_0)$.

The error is below 1% of L_z. This small deviation is compensated by slightly increasing the bubble spacing in $z-$direction.

The complete setup is shown in Figure 5.4. Table 5.1 gives geometrical and numerical parameters of the computation.

Two of these close-packed layers of fixed spheres are placed in the upper part of

Table 5.1: Parameters for the detailed investigation of the FCC preference.

Quantity	Symbol	Value
Domain size	$L_x \times L_y \times L_z$	$8D \times 16D \times 8D$
Bubble diameter [m]	D	0.002
Number of fixed Bubbles [-]		112
Number of mobile Bubbles [-]	N_b	250 or 1
Material properties	Aluminium	Table 1.1
Spatial resolution [m]	Δs	1.25×10^{-4}
Time step [s]	Δt	10^{-4}
Lagrange points	n_L	806
Drainage velocity [m/s]	v_0	$0 \ldots 0.05$

the domain, triggering crystal growth and allowing for easy characterization of the generated crystal. Below, 250 mobile bubbles are released successively at randomly chosen points. The basic collision model is applied.

5.3   Quantification of crystalline order

The mobile bubbles rise and agglomerate below the existing crystal, choosing an FCC or an HCP position or sometimes remain unordered. For the third and fourth layer one can characterize each bubble distinctly as HCP or FCC from its horizontal position relative to the fixed layers, as explained in Figure 5.5. Bubbles within $0.1\,D$ distance to a theoretical crystalline position are considered to be crystalline. This simulation has been carried out several times with different random initial positions of the mobile bubbles. The final positions of the mobile bubbles have been characterized. In that way, statistical data for the numbers n_{FCC}, n_{HCP} of bubbles in FCC and in HCP arrangement, respectively, have been derived. The preference of FCC P_{FCC} is characterized by

$$P_{FCC} = \frac{n_{FCC} - n_{HCP}}{n_{FCC} + n_{HCP}}. \tag{5.1}$$

Figure 5.6 shows the preference P_{FCC} for different drainage velocities, derived from 30 computational runs each. The error bar indicates the statistical uncertainty of the mean value. Obviously, there is no FCC preference in the third layer, but a clear preference in the fourth layer. This result is confusing at this point, but will be explained below. Figure 5.6 also shows, that the FCC preference increases with increasing drainage velocity. This seems to support the theory, mentioned above, that the hydrodynamic interactions cause the FCC preference.

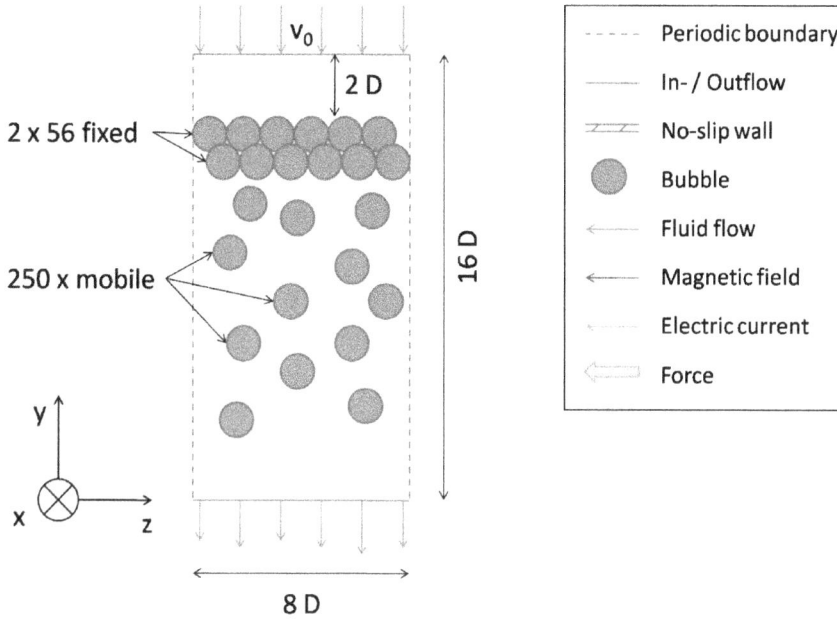

Figure 5.4: Numerical setup for the statistical investigation of the preference of FCC over HCP packing.

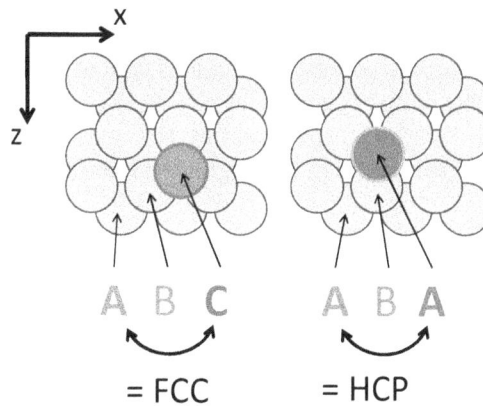

Figure 5.5: Algorithm of the determination of the packing character of a bubble. If the x-z-position corresponds to a bubble in the second-next layer above, it is an HCP bubble, if it corresponds to an empty position, it is an FCC bubble.

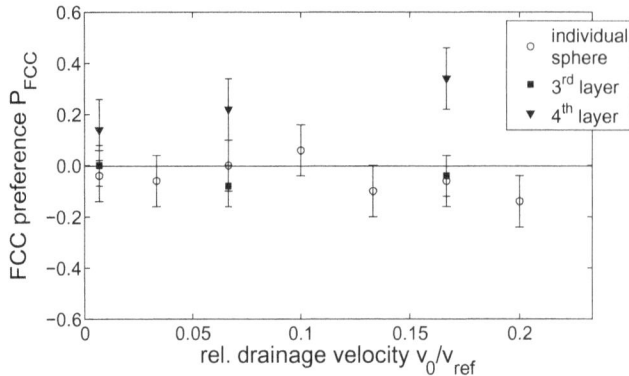

Figure 5.6: Preference of FCC over HCP packing according to Equation (5.1). Preference is shown as a function of the drainage velocity v_0 for the third layer, the fourth layer and an individual bubble.

5.4 Hydrodynamic effects in bubble agglomeration

In order to extract the hydrodynamic effects, a second setup is developed. This second setup is similar to the first one, shown in Figure 5.4, but instead of 250 mobile bubbles only a single one is placed below the fixed layers. The single bubble rises and agglomerates either in FCC or in HCP arrangement, as described by Figure 5.5. Then, the simulation is started over again with the mobile bubble positioned at another, random horizontal position at the bottom of the domain. This has been repeated for different drainage velocities, 100 times each. The preference P_{FCC} for this single bubble is also plotted in Figure 5.6. Obviously, the single bubble does not prefer one packing over the other. This means, the FCC preference cannot be a purely hydrodynamic feature as suggested by the hydrodynamic theory. Instead, interaction of many mobile bubbles in the agglomeration process has to play a crucial role.

5.5 Mechanical stability

The author develops a different explanation, taking the mechanical stability of the packing structure into account. In FCC the bubbles are situated along a straight line inclined with respect to the vertical direction (dash-dotted line in Figure 5.7 bottom left). Consequently, disturbing forces from consecutively impacting bubbles are transferred through the crystal along these straight force lines, which does not destroy the cluster. In an HCP arrangement, however, no such unbroken lines exist as illustrated by Figure 5.7 bottom right. Force transfer through incomplete

HCP packing can result in residual forces, pointing outward, that eject bubbles from their crystalline position and consequently, destroy the already existing HCP crystal. These destruction events in already formed HCP structure could indeed

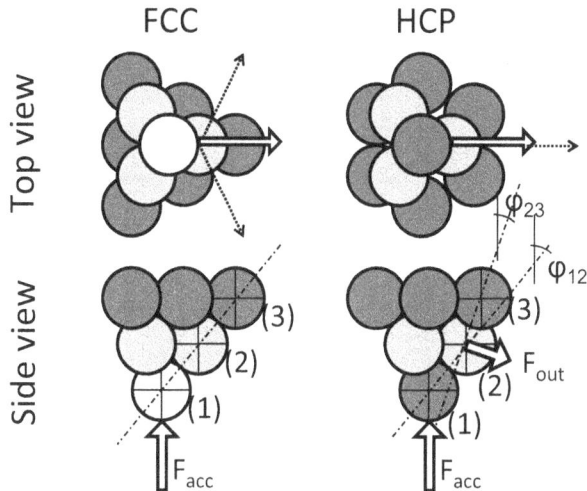

Figure 5.7: Mechanism of HCP instability. In FCC, an additional tip force F_{acc} is transferred along the straight line (1)-(2)-(3). In HCP, line (1)-(2) and line (2)-(3) differ in the angle ϕ. Force transfer causes an outward force that destabilizes HCP packing.

be observed in the simulations described above. To prove the appearance of this mechanism in a systematic way, the temporal evolution of HCP and FCC clusters is recorded. Figure 5.8(b) shows the number of bubbles in FCC and HCP arrangement in the third and fourth layer, respectively, averaged over 30 computations with $v_0/v_{ref} = 0.2$. The first and second layer, counted from top to bottom, consist of fixed bubbles. At the same time, Figure 5.8(a) documents the bubble distribution at three discrete times, showing the filling state of the layers. The reference time $t_{ref} = D/v_{ref}$ is the ratio of bubble diameter and reference velocity. At $t/t_{ref} \approx 5$ the filling of the third layer starts. Both arrangements, FCC and HCP grow with the same speed and consequently end up with nearly the same fraction. At $t/t_{ref} \approx 10$ the filling of the fourth layer starts, the third layer is not yet filled completely. Until $t/t_{ref} \approx 12$, when the third layer is filled, one can see in the fourth layer, that the number of FCC grows faster while the number of HCP bubbles nearly stagnates. This difference is caused by the destruction of already formed HCP arrangements. While a bubble in FCC arrangement in the fourth layer remains stable, an HCP bubble is likely to displace a bubble in the still incomplete third layer, which makes the HCP bubble leaving its position in the fourth layer. Consequently, FCC has the lead in filling the fourth layer. This lead is very small, but one bubble in FCC and HCP position, respectively, triggers growth of the same arrangement in its surrounding, because it blocks three positions of the other type. Therefore, in the period $t/t_{ref} \approx 13 \ldots 20$, FCC grows faster than HCP, amplifying the FCC preference.

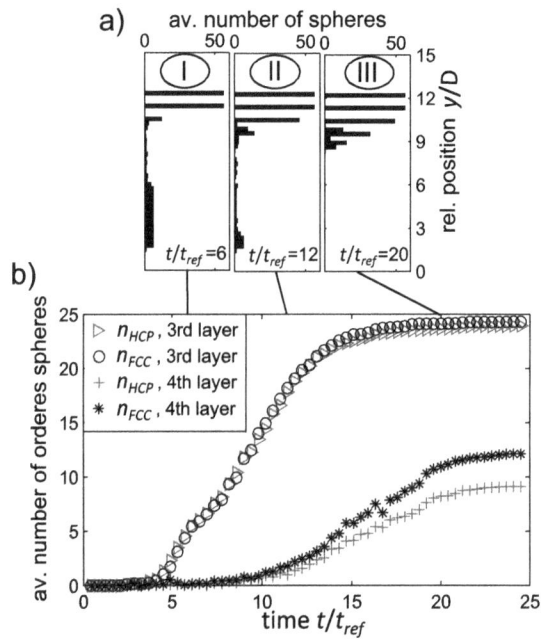

Figure 5.8: Average height distribution of the mobile bubbles at three different times (a) and temporal evolution of the number of bubbles in HCP and FCC arrangement in the third and fourth layer, respectively (b).

Additionally, in the case of $v_0/v_{ref} = 0.2$, all events when a bubble in the third layer is ejected from a crystalline position, are analysed. The results are shown in Figure 5.9, averaged over 30 simulations. Figure 5.9(a) shows the angle Φ_e in

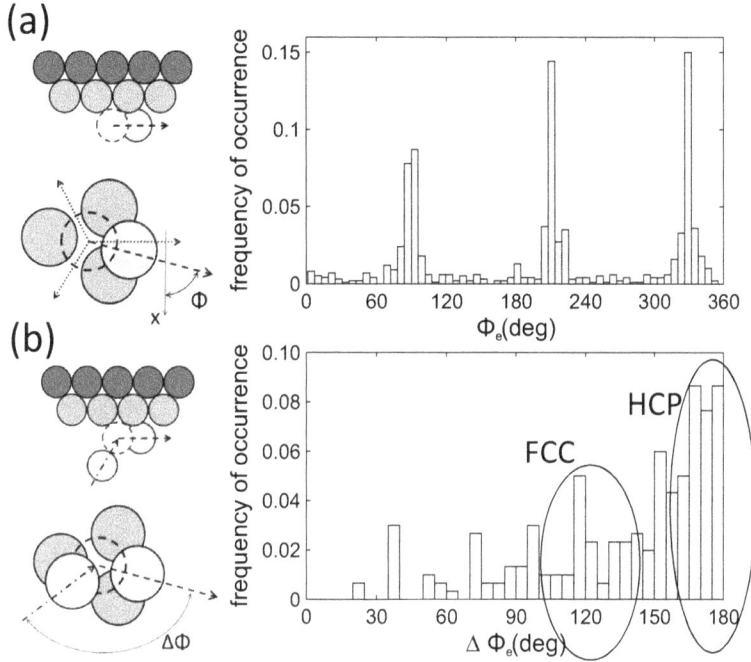

Figure 5.9: Analysis of ejection events. If a bubble in the third layer is ejected from a crystalline position it is very likely to move along a valley in the second layer, corresponding to an angle Φ_e in the x-z-plane of 90°, 210° or 330° (a). This event is more likely to be caused by an HCP bubble, corresponding to $\Delta\Phi_e = 180°$ than by an FCC bubble, corresponding to $\Delta\Phi_e = 120°$ (b).

the horizontal plane, under which a bubble in the third layer is ejected from its crystalline position. One can see, that it very likely moves through one of the three valleys in the second layer above. Figure 5.9(b) reports on events, when the bubble in the third layer is ejected due to the repelling force from a bubble in the fourth layer and when the fourth-layer bubble replaces the third-layer bubble. These events represent the instability mechanism of an HCP pyramid, as explained above. The angle $\Delta\Phi_e$ in Figure 5.9(b) is the angle in the horizontal plane between ejection direction and repelling force direction. In case of the disintegration of an HCP pyramid, according to the mechanism presented above, $\Delta\Phi_e$ equals about 180°. In case of the disintegration of an FCC pyramid, it equals about 120°. One can see in Figure 5.9(b) that disintegration of HCP happens more often than disintegration of FCC. This supports the hypothesis above, that HCP in the fourth layer disintegrates more likely than FCC.

The stability argument also explains, why there is no FCC preference in the third layer. An HCP bubble can only destroy the packing if the layer above is mobile.

This is the case for the fourth, but not for the third layer. Thus, only the fourth layer can show an FCC preference. And this argument explains, why the FCC preference is more significant at higher drainage velocities. A drainage flow assists in destabilizing the packing. Thus, with higher drainage flow, the packing is more likely to disintegrate. To prove this, a stability map for a small upside-down pyramid is derived from several simulations. The corresponding setup is shown in Figure 5.10

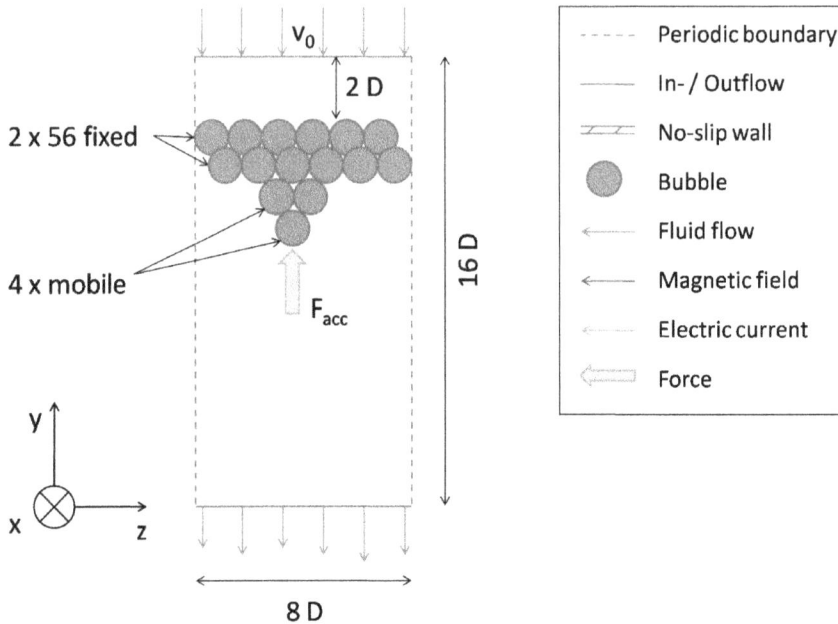

Figure 5.10: Numerical setup for the investigation of the stability of an upside-down pyramid exposed to drainage v_0 and tip force F_{acc}.

A mobile upside-down pyramid of FCC or HCP, as shown in Figure 5.7 is placed below two layers of fixed bubbles. An additional force F_{acc} on the tip bubble represents the impact of another arriving bubble. Also a drainage flow has been applied. Simulating different combinations of tip force and drainage flow, it is tested, whether the pyramid remains stable or disintegrates. Figure 5.11 shows the resulting stability map with F_{acc} normalized by the buoyancy force F_b. Obviously, there is a large region of conditions under which FCC is stable while HCP is unstable. With increasing drainage velocity, the amount of additional tip force F_{acc} that can be tolerated decreases.

5.6 Experimental reproduction

In order to support the numerical results, the stability test for an upside-down bubble pyramid is carried out experimentally. Using a microfluidic device [85],

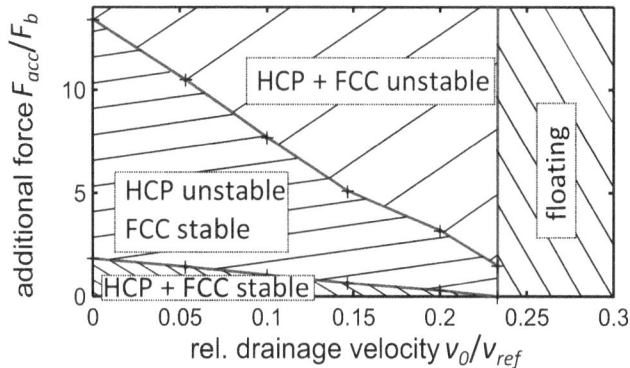

Figure 5.11: Stability map for an upside-down pyramid in FCC and HCP arrangement. Above $v_0/v_{ref} = 0.23$ both pyramids start to float and disintegrate. Below $v_0/v_{ref} = 0.23$ FCC can resist to a higher tip force F_{acc} than HCP.

small bubbles are placed below a graved top plane. The grave in the top plane is either a hexagon or a triangle, forcing the growth of an HCP or an FCC pyramid, respectively. With a high speed camera, the behaviour of the pyramid is observed. As shown in Figure 5.12, the FCC pyramid remains stable while the HCP pyramid is disintegrated according to the instability mechanism, presented above. This provides further evidence, that an incomplete FCC structure is indeed more stable than an incomplete HCP structure. Videos of this experiment are available as supplemental material to [94].

5.7   Conclusions

In this Chapter, a stability argument is formulated that is, at least in our simulations, at the origin of the preference of FCC over HCP bubble packing. Due to the generality and simplicity of this mechanism it is most likely also applicable in other systems. This principle of higher stability is well known by fruit merchants, arranging their goods in FCC and not in HCP ordering, shown in Figure 5.13. In an FCC packing, all contact points of spheres are arranged on straight lines. So FCC consists of sphere columns in [110], [1$\bar{1}$0], [011], [01$\bar{1}$], [101] and [10$\bar{1}$] direction. Consequently, forces, acting in any direction are decomposed into principal forces acting along these straight columns, leading automatically to an equilibrium of forces at every sphere and thus, the cluster remains stable. On the contrary, HCP structure does only contain three straight column of spheres in [01.0], [10.0] and [$\bar{1}\bar{1}$.0] direction, all situated in one layer. Thus, force that is transferred vertically to this layer through an HCP cluster is redirected at every sphere, causing a disequilibrium of forces on the spheres that tend to rearrange HCP packing. Figure 5.14 depicts this idea in a two-dimensional sense.

As proven experimentally with small bubbles and with metal spheres, this differ-

60

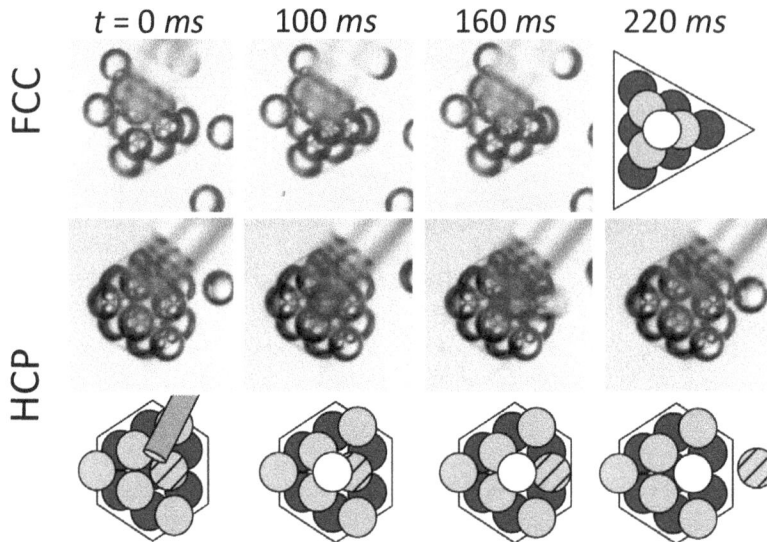

Figure 5.12: Experimental observation of the instability of HCP. An HCP arranged upside-down pyramid of small bubbles $D = 9\,$mm rearranges by its own buoyancy according to the described instability mechanism, while a FCC pyramid remains stable.

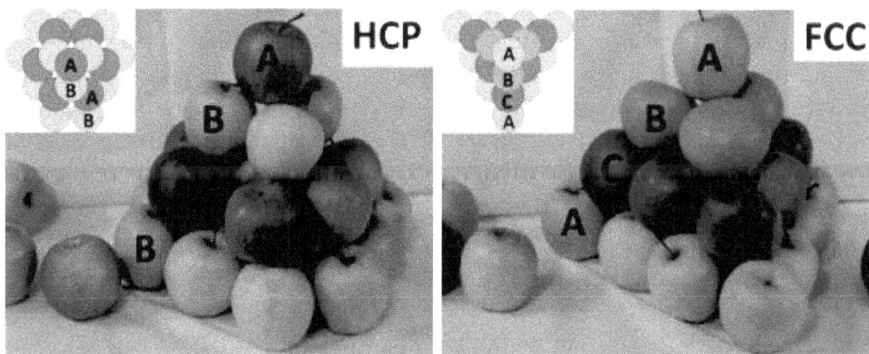

Figure 5.13: Arrangement of fruits in HCP and FCC ordering.

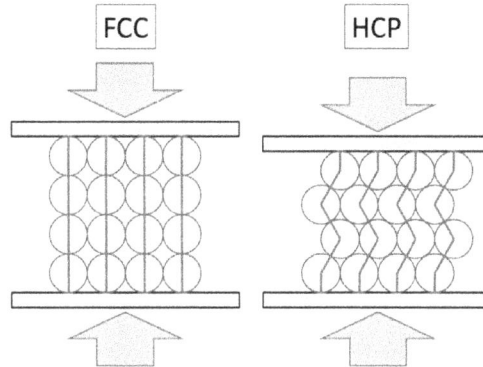

Figure 5.14: Idea of force transfer in FCC and HCP (geometrically incorrect!). In FCC force is transferred along straight lines, leading to an unstable equilibrium. In HCP, multiple changes of direction appear, resulting in a disequilibrium.

ence in force transfer causes an HCP pyramid to collapse while an FCC pyramid remains stable. Analysing the rearrangement events in the simulations, it is found that much more rearrangements of HCP elements than of FCC elements take place. One can conclude, that FCC is not the preferred, but the more stable arrangement for bubbles. Keeping the formation of metal foam in mind, this raises two questions. First, has FCC metal foam advantageous or disadvantageous properties compared to HCP? And second, is there a way of amplifying or inhibiting the preference of FCC over HCP packing?

6 Mechanical properties of solid foam

6.1 Motivation

This section bases on the manuscript [103] and on the Diploma thesis of Titscher [104]. As shown in Chapter 5, monodisperse bubbles tend to agglomerate in crystalline ordering. They prefer close-packed structures, such as FCC or HCP. Polydispersity of the bubbles prevent this ordering, resulting in the formation of random close-packed (RCP) bubble arrangements. This raises the question, whether one of these structures is particularly advantageous or disadvantageous for the mechanical properties of the generated solid foam. If such a structure could be identified, the goal of the manipulation of metal foam with electromagnetic fields would be to amplify or to avoid the generation of this structure.

In general, solid foams show a very rich, non-linear mechanical behaviour, including plastic deformation, buckling or rupture. For light-weight construction the focus is on elastic behaviour, because plastic deformation will cause permanent modification of the device, which must be avoided. To generate results that are applicable to different materials, independent from material properties such as the yield point, the focus in this work is on the linear elastic behaviour, corresponding to an infinitely small strain. To calculate the elasticity of different foam structures, Finite-Element simulations are carried out using ANSYS FEM. To do so, the foam is simplified by several assumptions. The shape of the bubbles is assumed to be spherical. Also, bubbles are separated, corresponding to closed-cell foam. The contribution of the trapped gas to the mechanical behaviour is neglected. The interstitial, solid material is supposed to be homogeneous. The mechanical influence of surfactants is also neglected.

The complement problem, the elastic properties of crystalline arrangements of spheres have been investigated experimentally and numerically by several authors [105–111]. The elastic properties of the interstitial material of these packings, however, has not yet been investigated sufficiently and comparatively.

Before computers made their breakthrough in science, the elastic properties of void material have been estimated by superpositioning the weakening of the material by a single void [112–114]. These analytical methods yield good results for low void fraction. With increasing void fraction, however, higher orders of interaction between the voids have to be taken into account [115–117]. Christensen [118] compared different micro-mechanic models available at that time. A high order model of this type was recently offered by Cohen et al. [119].

In 1992, Day et al. [120] developed a simple Finite Element Method (FEM) to calculate numerically the elastic properties of a two-dimensional material with circular voids. They investigated the influence of void fraction and topology separately and devised a simple analytical explanation for the calculated values. After 1992, increasing computer power became available for many research groups, resulting in further direct numerical simulations of the interstitial material of sphere or bubble arrangements [121–124].

The agreement between analytical methods [114, 119] and numerical methods [122] is very good. However, as is shown in Figure 6.5 below, the Young's modulus only depends weakly on the structure. Thus, small deviations between the graphs leave one puzzled, if a difference results from the uncertainty of the method or actually from the structural differences of the investigated material. In order to extract the structural effects, one therefore needs to apply an identical numerical method to different structures, taking much care of the numerical uncertainty. In this chapter, a comparative study of material with dense FCC, HCP and RCP packings of spherical voids is carried out, revealing the influence of the structure on the elastic properties. For comparison, also simple cubic (SC) and body-centred cubic (BCC) arrangements are taken into account

6.2   Definition of sphere structures

With the assumptions stated above, metal foam is only characterised by the mechanical properties of the solid material, the arrangement of the spherical voids and the void fraction. The void fraction Φ is the ratio between the volume of the spheres and the total volume. For a crystalline arrangement it depends on the sphere diameter D, the lattice spacing L_l, and on the maximum packing density $\Phi_{v,max}$ of the considered structure

$$\Phi = \Phi_{v,max} \left(\frac{D}{L_l} \right)^3 - \Phi_{v,max} \left(1 - \frac{l_l}{L_l} \right)^3. \qquad (0.1)$$

Defining the thickness of a lamella to be $l_l = L_l - D$ as displayed in Figure 6.1 yields the second equality in Equation (6.1). From the different structures mentioned above, cuboidic or cubic representative volume elements (RVE) are derived which are shown in Figure 6.2. Parameters of the chosen RVE are given in Table 6.1. Please note that for FCC two different RVE are applied and compared. The cubic RVE, labelled FCC, is a cube, bounded by planes in (100), (010) and (001). The hexagonal RVE, labelled FCCh, is a cuboid, bounded by (111), (1$\bar{1}$0) and (11$\bar{2}$) planes. This provides an additional test of the method applied by comparing the Young's moduli of the different RVE of the same structure. This is explained in more

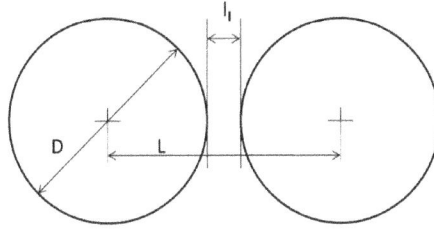

Figure 6.1: Geometry of sphere packing.

detail in Section 6.5. The RCP structure is special, since it does not correspond to

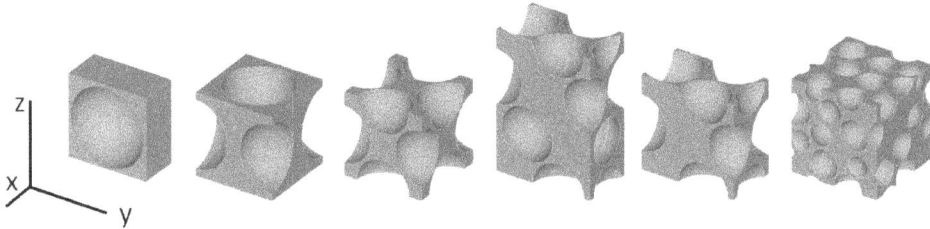

Figure 6.2: Sketch of the chosen RVEs, from left to right: Simple cubic (SC), Body-centred cubic (BCC), Face-centred cubic (FCC), Hexagonal Face-centred cubic (FCCh), Hexagonally close-packed (HCP), Random distributed (RCP).

a crystalline lattice. The sphere positions for this case are generated, using a gas-dynamic algorithm that is freely available [125]. Drugan et al. [126, 127] found that with six spheres in an RVE of disordered voids, the statistical uncertainty of the mechanical properties is below 5 %. Aiming for very high accuracy, here RVEs with 30 spheres are employed. The statistical uncertainty of this type of RVE is investigated, as reported in Section 6.5.

Table 6.1: Parameters of the RVEs of the structures considered. The size of the RVE in x-, y- and z- direction is denoted L_x, L_y, and L_z, respectively, n_b the number of spheres or bubbles in each RVE and $\Phi_{v,\max}$ the maximum void fraction of sphere packing.

Structure	Label	$(L_x \times L_y \times L_z)/L_1$	n_b	$\Phi_{v,\max}$
simple cubic	SC	$1 \times 1 \times 1$	1	$\frac{1}{6}\pi \approx 52,4\%$
body-centred cubic	BCC	$\frac{2}{\sqrt{3}} \times \frac{2}{\sqrt{3}} \times \frac{2}{\sqrt{3}}$	2	$\frac{\sqrt{3}}{8}\pi \approx 68\%$
face-centred cubic	FCC	$\sqrt{2} \times \sqrt{2} \times \sqrt{2}$	4	$\frac{\sqrt{2}}{6}\pi \approx 74\%$
hexagonal fcc	FCCh	$1 \times \sqrt{3} \times 3\sqrt{\frac{2}{3}}$	6	$\frac{\sqrt{2}}{6}\pi \approx 74\%$
hexagonal close-packed	HCP	$1 \times \sqrt{3} \times 2\sqrt{\frac{2}{3}}$	4	$\frac{\sqrt{2}}{6}\pi \approx 74\%$
random	RCP	$\approx 3.1 \times 3.1 \times 3.1$	30	$\approx 62\%$

6.3 Homogenisation and elastic properties

The idea of homogenisation is to find an equivalent homogeneous continuum material (EHC) that behaves, in a volume-averaged sense, mechanically similar to the heterogeneous RVE [128]. The mechanical behaviour is described by Hook's law, which can be written as

$$\langle \sigma \rangle = \mathbf{C} \langle \varepsilon \rangle , \tag{6.2}$$

with $\langle \ldots \rangle$ designating the volume average over the RVE with volume V. This law links the applied stress σ, the resulting strain ε and the stiffness matrix $\mathbf{C}$ [129]. For a valid EHC it is imperative that RVE and EHC store the same elastic energy E_{el} when applying the same global strain

$$\frac{E_{\mathrm{el}}}{V} = \frac{1}{2} \langle \varepsilon \rangle^T \mathbf{C} \langle \varepsilon \rangle . \tag{6.3}$$

The strain vector ε consists of 6 elements, which are normal strain ε_{xx}, ε_{yy}, ε_{zz} and shear strain ε_{xy}, ε_{yz} and ε_{zx}. The stress vector σ contains 6 respective elements. Due to symmetry, the 6×6 elements of $\mathbf{C}$ consist of 21 independent elements. Starting from a heterogeneous RVE, these can be computed by applying 21 independent load cases and calculating the corresponding elastic energy. These load cases are defined by 21 sets of strain ε.

The diagonal elements C_{kk} result from sets of strain with only one element $\varepsilon_{kk} \neq 0$

$$C_{kk} = 2 \frac{E_{\mathrm{el},kk}}{\varepsilon_{kk}^2 \, V} . \tag{6.4}$$

The off-diagonal elements C_{kl} result from sets with two strains $\varepsilon_k \neq 0$, $\varepsilon_l \neq 0$

$$C_{kl} = \frac{E_{\mathrm{el},kl} - E_{\mathrm{el},kk} - E_{\mathrm{el},ll}}{\varepsilon_k \, \varepsilon_l \, V} . \tag{6.5}$$

Note that $C_{kl} = -C_{lk}$ holds.

From the stiffness matrix C_{kl} one obtains the compliance matrix $\mathbf{D} = \mathbf{C}^{-1}$ by calculating its inverse. And finally, one obtains the Young's modulus E_i of the RVE by analysing the first three main diagonal elements of the compliance matrix

$$E_x = \frac{1}{D_{11}}, \quad E_y = \frac{1}{D_{22}}, \quad E_z = \frac{1}{D_{33}} . \tag{6.6}$$

The Young's modulus of the void structure, normalised by the Young's modulus of the matrix material E_0, is the main material parameter considered here. In some cases, a mean Young's modulus is analysed, defined as $E_{\mathrm{m}} = 1/3(E_x + E_y + E_z)$.

The influence of the Poisson ratio ν_{m} of the matrix material on the Young's modulus of the void material generally is not negligible. In the present study, $\nu_{\mathrm{m}} = 0.4$ is used. Only in rare cases to allow comparison with literature ν_{m} is adjusted.

Voigt's rule of mixture [130] provides an upper limit E_{Voigt} for the Young's modulus of a porous material

$$\frac{E}{E_0} \leq (1 - \Phi) = \Phi_{\mathrm{s}} = \frac{E_{\mathrm{Voigt}}}{E_0} , \tag{6.7}$$

with Φ_s the fraction of solid material. By comparing the actual Young's modulus with the upper limit, one can estimate the efficiency of material usage in the referring structure.

6.4   The Finite-Element method

To apply a certain strain and to calculate the resulting elastic energy E_{el}, the commercial software ANSYS FEM is used. The RVE's are meshed with tetrahedral elements with quadratic ansatz functions. The mesh parameter N_g, which is the number of elements over a distance L_1, quantifies the resolution of a particular grid. The minimum lamella thickness that is investigated is $l_1 = 0.05\, L_1$ which corresponds to the size of one finite element at $N_g = 20$. In order to impose zero-strain or zero-stress periodic boundary conditions, it is necessary to guarantee identical grids on opposing faces. The periodic displacement is then realised by adding additional restricting equations to periodic point pairs.

6.5   Validation

The results presented below show that differences in the mechanical behaviour occur if voids are arranged in different ways, but that these differences are small. In this situation it is important to assess the accuracy of the method applied and to demonstrate that the uncertainty of the results is below the differences addressed. In order to derive the uncertainty of the method three ways of validation are applied. The first method is a grid study. With increasing resolution, a numerical solution should converge toward the exact solution. However, since the resolution is usually limited due to limited computer power, one has to choose a resolution which yields results with sufficiently small deviation from the exact solution. For the grid study, a cubic representative, FCC, and a hexagonal representative, HCP, with thin lamellas $l_1/L_1 = 0.05$ are computed, corresponding to a void fraction of 0.65. Since there is no analytical or exact solution available for this problem, it is computed with different resolutions $20 \leq N_g \leq 44$ and the resulting Young's modulus fitted with a power function. The results are shown in Figure 6.3. The method is third order accurate in $x-$ and $z-$ direction, but only second order in $y-$ direction. The different order might correspond to the orientation of the lamellas, since these are the critical regions in terms of grid resolution. According to Figure 6.3, the numerical error due to resolution is below 1% for $N_g = 20$ and below 0.2% for $N_g = 40$. The results, reported below, are obtained with $N_g = 20$, except if stated otherwise.

The second validation method is the comparison with literature [114, 119, 121, 122]. To this end, the Poisson ratio ν_m is adjusted to meet the literature values. Figure 6.4 demonstrates the good agreement between literature data and results of the present method. In the case of the RCP structure, simulation of 45 different, randomly generated RVE's are performed and statistically analysed. Figure 6.4, (bottom) shows the histogram of the results, the standard deviation of the Young's modulus

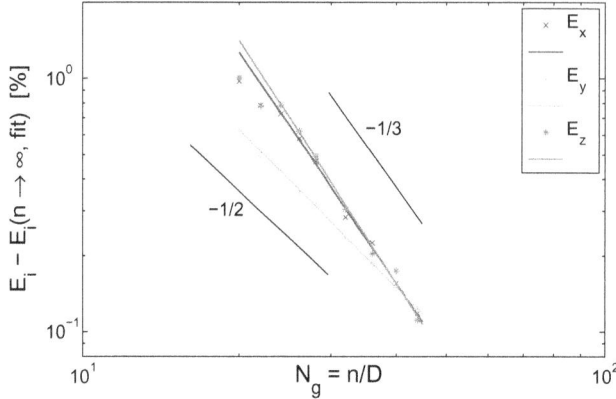

Figure 6.3: Dependency of the numerical deviation of the Young's modulus on the grid resolution N_g. The assumed exact value $E_i(n \rightarrow \infty)$ is derived from fitting a power function. The lines represent linear fitting of the data on a logarithmic scale.

and the confidence interval for the mean value. The confidence interval measured equals 1%, which is in the same order of magnitude as the numerical uncertainty of the crystalline simulations. The mean Young's modulus is in good agreement with [121].

The third method of validation is a sensitivity test. As pointed out in Table 6.1, the FCC structure is calculated using two RVE with different orientations. The resulting Young's moduli are different due to their dependency on orientation. But by applying a unitary transformation

$$\mathbf{C}'_{\mathrm{FCC}} = \mathbf{K} \, \mathbf{C}_{\mathrm{FCC}} \, \mathbf{K}^T. \tag{6.8}$$

to the stiffness matrix $\mathbf{C}$, the computed result can be transformed into the same coordinate system. In theory, $\mathbf{C}'_{\mathrm{FCC}}$ and $\mathbf{C}_{\mathrm{FCCh}}$ should be equal. But due to computational uncertainty, they show a small deviation, as shown in Table 6.2. From this deviation one can derive another estimation of the uncertainty of the computation. In the present case, the uncertainty is below 0.6%. This value is in good agreement with the numerical uncertainty, derived from the grid study.

Table 6.2: Comparison of the Young's modulus, calculated directly for FCCh or by transformation of FCC, both obtained with a resolution of $N_g = 20$.

Young's modulus	FCCh, direct	FCC, transformed	Relative deviation
E_x/E_0	0.1847	0.1858	0.5941%
E_y/E_0	0.1862	0.1858	0.1912%
E_z/E_0	0.1974	0.1975	0.0702%

Taking into account the different tests performed on the uncertainty of the method we can therefore conclude that overall the uncertainty of the employed method for the computation of Young's modulus is less than 1%.

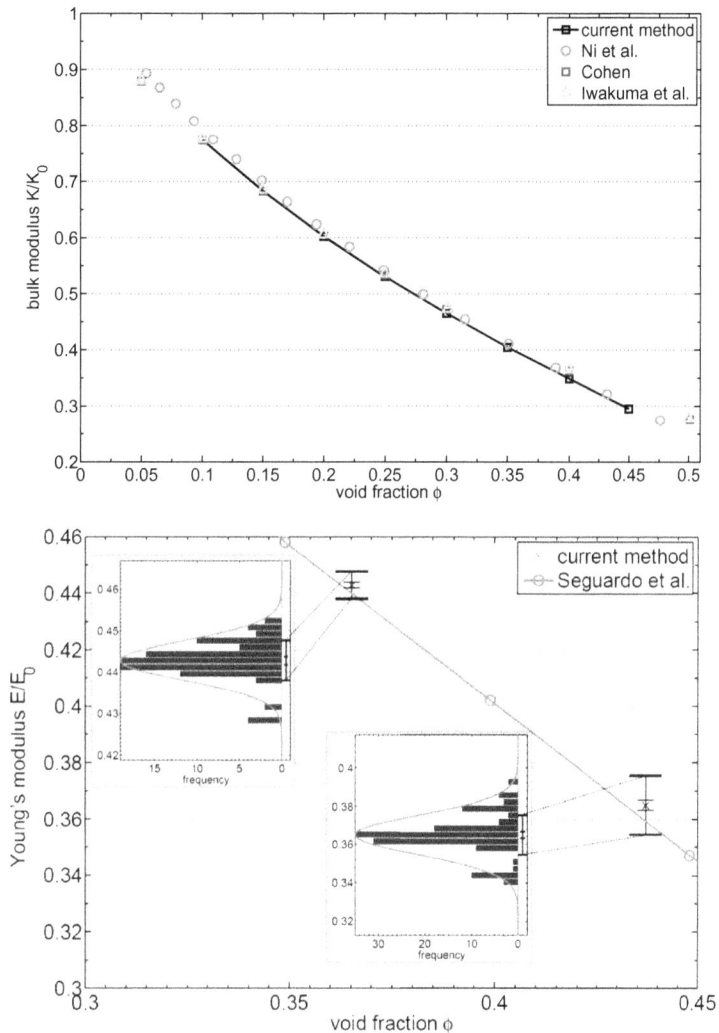

Figure 6.4: Comparison of the present method with literature data. Simple cubic with $\nu_m = 0.3$ [114, 119, 122] (top) and random distribution with $\nu_m = 0.25$ [121] (bottom) show good agreement.

6.6   General comparison of Young's moduli

The Young's modulus for different configurations, given in Table 6.1, is computed for different void fractions Φ. The void fraction of the RVE is varied by changing the sphere diameter D whithout changing the bubble centre positions. The mean values of the Young's modulus E_m are plotted in Figure 6.5 (top). Generally, the mean Young's moduli of different structures appear to be relatively close to each other. The only exception is SC, which is in $x-$, $y-$ and $z-$ direction a bit stiffer than the others. The reason for this is the structure of the interstitial material. It consists of straight columns, supporting the load very effectively. This can be seen in Figure 6.6. Please note that in this type of Figure the sphere diameter is artificially increased, in order to create a transparent structure.

6.7   Comparison of cubic structures

In order to distinguish more clearly the other structures, the data from Figure 6.5 (a) is replotted in a different style in Figure 6.5 (b). Voigt's law (6.7) provides an upper limit for the Young's modulus for any porous material by a simple rule of mixture. Figure 6.5 (b) displays the relative deficiency to this upper limit

$$\frac{E(\Phi) - E_{\text{Voigt}}(\Phi)}{E_0} = \frac{E(\Phi)}{E_0} - \Phi_s, \tag{6.9}$$

for each structure, void fraction and direction. In case of BCC, FCC and RCP, the Young's modulus in $x-$, $y-$ and $z-$ direction has to be equal for reasons of symmetry. Thus, only the mean value is plotted to avoid overloading the figure. The error bars for the crystalline structures are derived from the grid study mentioned above. The error bars for the RCP case correspond to the confidence interval of the mean value.

BCC and FCC both have a very low Young's modulus. The reason is that they do not consist of column-like arranged material in $x-$, $y-$ or $z-$direction, as shown in Figure 6.6. Also in RCP, there are no straight columns, which yields a lower Young's modulus for the same reason.

6.8   Comparison of hexagonal structures

Due to their nearly identical structure, HCP and FCCh show very similar Young's moduli. The values in $x-$ and $y-$ direction differ only by 3%, which is close to the numerical error. Thus, to distinguish these structures, the case of $\Phi = 0.65$ is computed with a higher resolution, $N_g = 40$, which yields numerical errors below 0.2 %. The resulting Young's moduli are given in Figure 6.7. With increased resolution the differences between HCP and FCCh remain, proving their physical

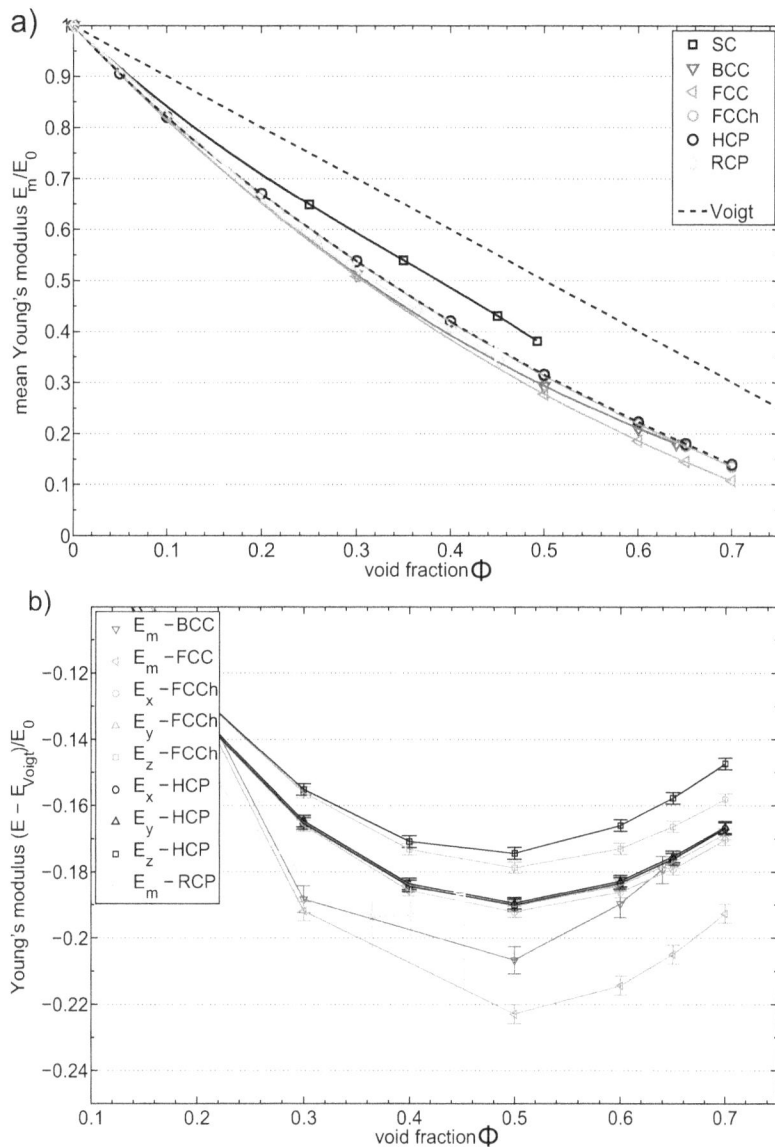

Figure 6.5: Normalised mean Young's modulus for a range of void fractions Φ for different arrangements of spherical voids. (a) Average Young's modulus E_m. (b) Normalised difference between the Young's modulus and the maximal Young's modulus E_{Voigt}.

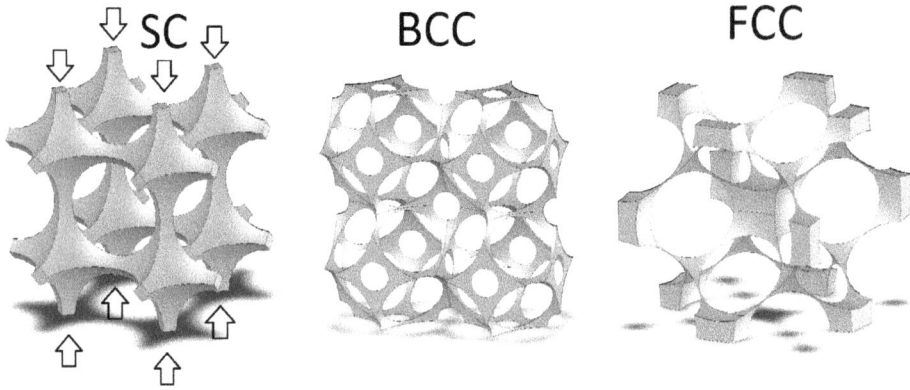

Figure 6.6: Visualisation of the structure of the interstitial material in SC, BCC and FCC structure, made apparent by artificial increasing of void diameter beyond overlap.

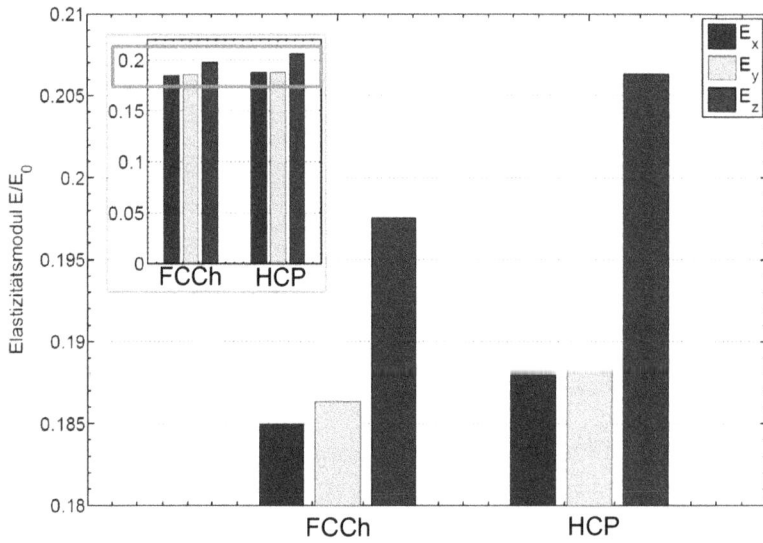

Figure 6.7: Comparison of the Young's moduli of FCCh and HCP for $\Phi = 0.65$, calculated with a high resolution of $N_g = 40$, yielding a numerical uncertainty of 0.2%.

origin. The difference in $z-$direction equals 5.4%, which are now discussed with the help of Figure 6.8. In $z-$direction, HCP offers one unbroken column "C" and two columns "A", broken in every second layer, which overall gives a relatively high Young's modulus. The force distribution obtained from the simulation (not reported here) shows that the "C"-column takes most of the load while the "A"-columns are very much unstressed. On the countrary, FCCh in $z-$direction has three columns "B" which are broken only in every third layer. Consequently, to transfer load in $z-$direction, it has to be transfered between the columns, which puts high shear stress on the thin lamellas. This results in an overall slightly lower Young's modulus than observed for HCP.

The difference between the Young's moduli of HCP and FCCh in the $x-$ and

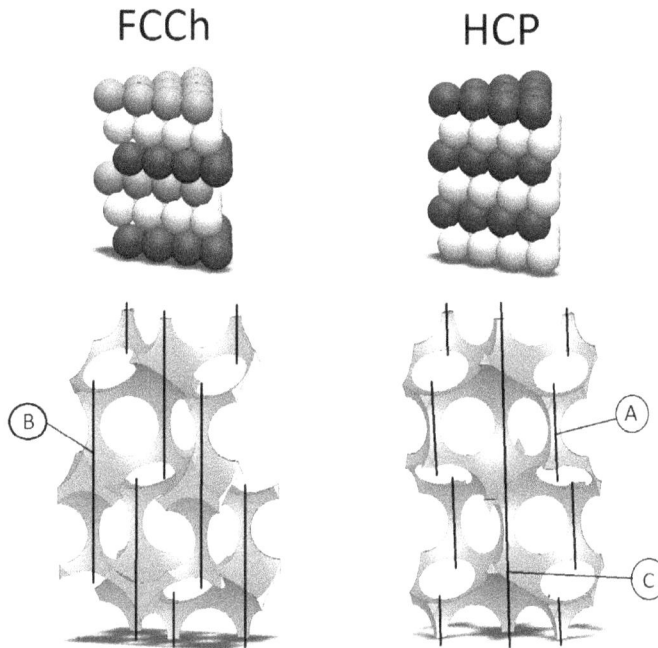

Figure 6.8: Visualisation of the column structure of the interstitial material obtained with spherical voids in FCCh structure (left) and HCP structure (right). Arrangement of voids (a) and Visualisation of interstitial material by artificial increasing of void diameter (b).

$y-$direction equals about 1%, which is now discussed with the help of Figure 6.9 nad 6.10. The interstitial material of HCP and FCCh packing consists of the same type of layers. But these layers are stacked and linked in a different way. Figure 6.10 sketches the mechanism, when these layers are compressed in $x-$ or $y-$ direction and relax in $z-$direction. In the relaxation, these layers tend to fold (Figure 6.10, left) which result in the shift of nodes upward and downward. In HCP (Figure 6.8, right), the layers are stacked in a way that creates links "A" and links "C". Under relaxation in $z-$direction, "A" is compressed from both sides, while "C" is stretched,

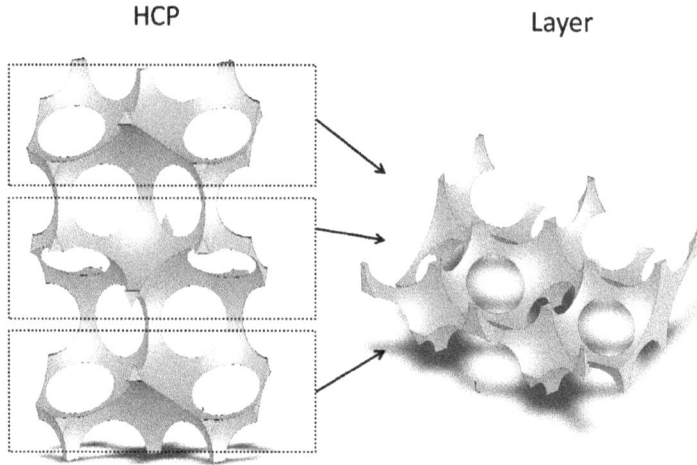

Figure 6.9: Visualisation of the interstitial material in an $x - y$ layer.

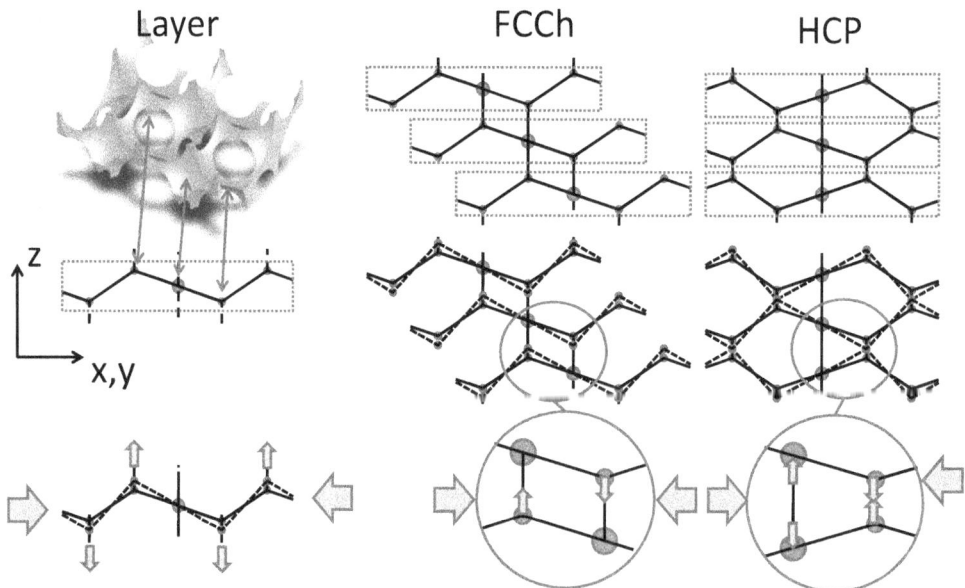

Figure 6.10: Sketch of the mechanism of relaxation of an $x - y$ HCP layer isolated (left), in FCCh (middle) and in HCP (right).

which generates additional tension and consequently hinders the relaxation process. On the contrary, in FCC (Figure 6.10, middle) only one type of link "B" exists (Figure 6.8, left). Under compression, all "B" relax parallel, which allows for easier relaxation and thus, a slightly lower Young's modulus in $x-$ and $y-$direction. The benefit of relaxation is different in $x-$ and $y-$direction, thus the Young's modulus of FCCh in $x-$ and $y-$direction is not the same. But, since this $z-$relaxation in the HCP structure is small, there is nearly no difference between $x-$ and $y-$ direction in this case.

6.9   Conclusions

The investigation of the elastic properties has revealed a small influence of the sphere arrangement on the Young's modulus of the resulting foam. Structures that contain straight columns of interstitial material, e.g. SC, are stiffer in the direction of these columns, since they transfer much of the applied load. Structures with broken columns or no columns are relatively soft, because force transfer relies on the shear resistance of the thin lamellas or bending resistance of the Plateau borders. Both are relatively low, especially for dry foam. Concerning the differences between HCP and FCCh, one has to state that the structural differences are small. The HCP arrangement is 5.4% stiffer in $z-$direction, since HCP offers an unbroken column in $z-$direction while all FCCh columns in this direction are broken. In the $x-$ and $y-$direction the differences are about 1% and thus, practically irrelevant. Compared to RCP at the same void fraction, FCCh and HCP both offer an about 10% higher Young's modulus. This might be desireable for light-weight-applications. But, one buys this higher stiffness with the consicerable effort of generating monodispersed bubbles.

From the author's point of view, the most significant and relevant difference is the difference between the stiffness at maximum void fraction of RCP on the one hand and FCCh and HCP on the other hand. The highest void fraction achievable in RCP equals about $\Phi = 0.63$, which yields a relative Young's modulus E_m/E_0 of about 0.33. In contrast, the highest void fraction of FCCh and HCP equals $\Phi = 0.74$, corresponding to a relative Young's modulus of approximately 0.1, which is by a factor of more than 3 softer than RCP. FCCh is here slightly more preferable, since it is even softer than HCP. This might be of particular interest for elastic devices or shock absorbing applications. This big difference might also justify the effort of generating monodisperse bubbles. Is has to be investigated whether the amount of crystalline ordering can be modified by adding electromagnetic fields.

7 Manipulation of foam with magnetic fields

7.1 Motivation

This chapter is based on the publication [61]. In Chapter 4 the relation between pressure drop and drainage velocity for unmodified bubble agglomeration has been investigated. This relation is important because it determines the rate of liquid metal that drains out of the foam. The pressure gradient that drives drainage is constant, defined by gravity and the liquid metal density. Depending on the drainage resistance of the foam, a volumetric flow of liquid metal results, as expressed in the drainage equation [131, 132]. To reduce this volumetric flow, the drainage resistance has to be increased. In this chapter, the influence of a horizontal magnetic field on drainage resistance is investigated. As in Section 4.3 above, the setup of static drainage is applied to guarantee steady conditions. This means, a constant drainage flow is introduced from top to bottom through the agglomerated bubbles. The pressure drop that is necessary to drive this flow is measured. The ratio of pressure drop to drainage velocity characterizes the drainage resistance.

In general, a magnetic field is well known to damp perpendicular movement of conducting liquid [21–26]. Assuming a magnetic field in negative z–direction $-B_0\mathbf{e}_z$ and a flow in negative y–direction $-v_0\mathbf{e}_y$, this results, for neglected electrical potential Φ_{el}, in an electric current

$$\mathbf{j} = \sigma_{\mathrm{el}}(\mathbf{u} \times \mathbf{B} - \nabla\Phi_e) = \sigma_{\mathrm{el}}B_0v_0\mathbf{e}_x. \tag{7.1}$$

The interaction of this current with the magnetic field causes a Lorentz force $\mathbf{f}_{\mathrm{L}}$

$$\mathbf{f}_{\mathrm{L}} = \mathbf{j} \times \mathbf{B} = \sigma B_0^2 v_0\mathbf{e}_y, \tag{7.2}$$

which depends quadratically on the magnetic field strength B_0 and acts in opposite direction of the flow velocity and therefore, damps the flow. The electric charge has to be conserved. Therefore, the electric current has to form a closed loop. This mechanism is well described and analytically solved in the so-called Hartmann flow [21]. This deals with a laminar flow in a parallel channel, as depicted in Figure 7.1(left). The positive fluid velocity throughout the channel generates a positive electric current at each point. In case of conducting walls, the electric current loop

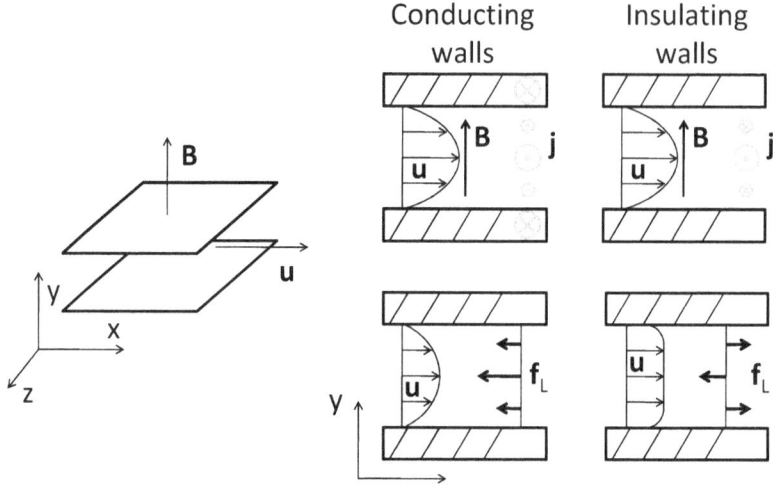

*Figure 7.1: Sketch of the mechanism of a Hartmann flow with and without conducting walls. (left) Sketch of the Hartmann flow configuration. (top) electric current **j** that is generated by the Interaction of the fluid velocity **u** and the magnetic field **B**. (bottom) Resulting Lorentz force distribution **f**$_L$ acting on the fluid.*

is closed through the side walls, as depicted in Figure 7.1(middle) This generates Lorentz forces that decelerate the flow. But, in case of insulating walls, the electric current cannot flow back through the walls. An electric potential gradient $\nabla\Phi_{el}$ in $z-$direction is generated, that causes a recirculation of the electric current, shown in Figure 7.1(right). This means, in regions of high flow velocity a positive current is induced, damping the flow, while in regions of low flow velocity a negative current balances the conservation of charge and accelerates the flow. In total, the magnetic field equalizes the differences in the flow velocity.

In this work, a drainage velocity in $-\mathbf{e}_y$ direction and a magnetic field in $-\mathbf{e}_z$ direction generate an electric current in $+\mathbf{e}_x$ direction. In x-direction periodic boundary conditions are applied, that do not force the electric current to form closed loops. The electric current would close the loop through the walls (far away). Therefore, the horizontal magnetic field is supposed to decelerate the drainage flow, which corresponds to an increase in drainage resistance. In the setup of static drainage this would correspond to an increase in the pressure drop since the drainage flow is kept constant. In this Chapter, the influence of a horizontal magnetic field on the pressure drop as well as on bubble mobility and bubble agglomeration is investigated.

7.2   Setup

In order to generate a steady state, the setup of constant drainage, as used in Chapter 4 is adapted. Figure 7.2 shows the arrangement. A constant drainage velocity v_0 channels downward through the computational domain. Boundary conditions in horizontal directions are periodic. At the top, 56 immobile bubbles are arranged

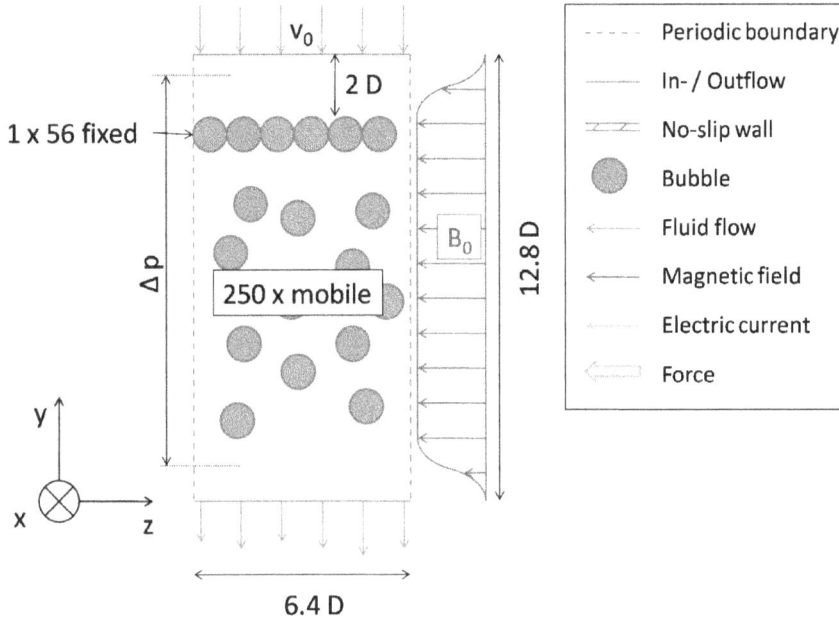

Figure 7.2: Numerical setup for the investigation of the influence of a horizontal magnetic field B_0 on the pressure drop Δp caused by drainage v_0.

in a regular hexagonal close-packed layer, similar to Chapter 5. Below, 250 mobile bubbles are randomly positioned and successively released one after the other starting with the highest ones, so that they rise and agglomerate below the fixed bubbles. The present investigations are carried out when all bubbles are released and a steady state for the pressure drop is reached.

A temporally constant horizontal magnetic field is imposed in the domain. Near the top and the bottom of the domain the magnetic field smoothly tends toward zero in order to get smooth in- and outflow conditions and is constant in space in between. At the top and bottom boundary the normal derivatives of the electric potential Φ_{el} are set to zero. The physical parameters of the setup are given in Table 7.1. A very high numerical resolution is applied, in order to resolve the fluid flow and the electric current within the interstitial liquid when bubbles agglomerate.

7.3   Results

To determine the total pressure drop Δp that is caused by the agglomerated bubbles the pressure difference between top and bottom of the domain is averaged in the horizontal cross-section and in time. The result is reported in Figure 7.3. The slope in the double-logarithmic plot corresponds to a linear relationship between pressure

Table 7.1: *Physical and numerical parameters of the computational setup.*

Quantity	Symbol	Value
Domain size	$L_x \times L_y \times L_z$	$6.4D \times 12.8D \times 6.4D$
Bubble diameter [m]	D	2×10^{-3}
Number of fixed Bubbles [-]		56
Number of mobile Bubbles [-]	N_b	250
Material properties	Aluminium	Table 1.1
Spatial resolution [m]	Δs	5×10^{-5}
Time step [s]	Δt	5×10^{-5}
Lagrange points	n_L	5024
Magnetic field strength [T]	B_0	$10^{-3} \ldots 10^{-1}$
Drainage velocity [m/s]	v_0	$10^{-3} \ldots 2 \times 10^{-2}$

drop and drainage velocity.

Two effects add to the pressure drop. On the one hand, there is a drainage resis-

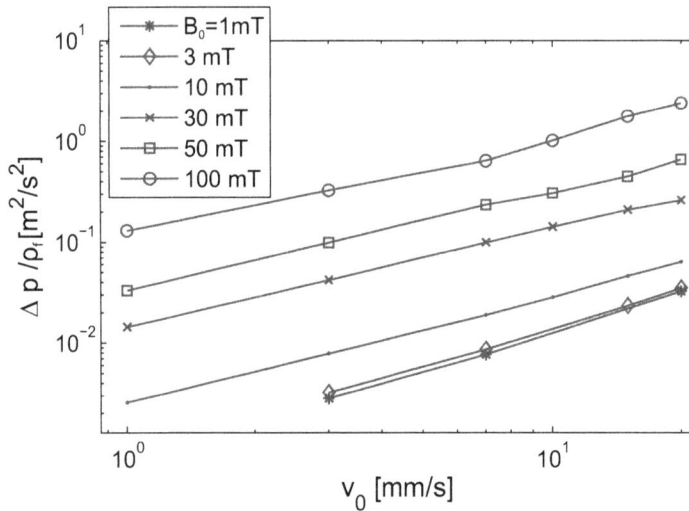

Figure 7.3: *Pressure drop versus drainage velocity for different magnetic field strengths.*

tance of the foam without magnetic field Δp_0 that is represented in Figure 7.3 by the data with negligibly small magnetic field B_0 of $1\,\mathrm{mT}$ and $3\,\mathrm{mT}$. On the other hand, the magnetic field causes an additional pressure drop that is also proportional to the drainage velocity. To extract the second effect, the unmodified drainage resistance Δp_0 (obtained from the case $B_0 = 1\,\mathrm{mT}$) is subtracted from the total pressure drop. Figure 7.4 shows the resulting pressure drop versus the strength of the magnetic field.

The slope in this graph is almost constant and corresponds to a quadratic relation-

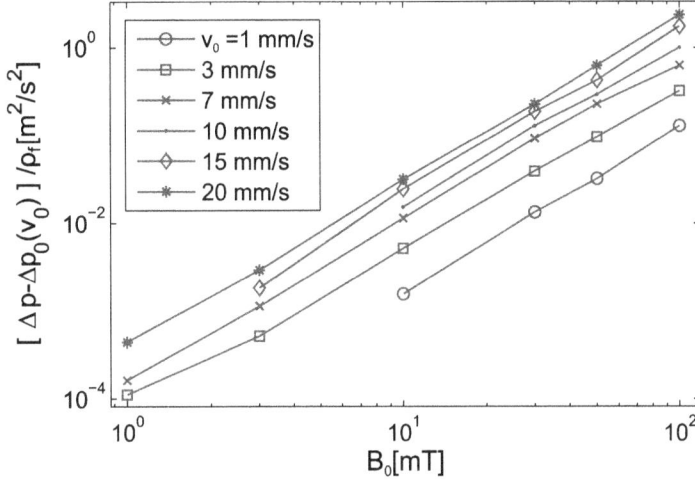

Figure 7.4: Increment of pressure drop versus strength of the magnetic field for different drainage velocities.

ship so for the total pressure loss the following ansatz is made:

$$\frac{\Delta p}{\rho_\mathrm{f}} = v_0(C_1 + C_2 B_0^2).$$ (7.3)

The values for C_1 and C_2 are determined by least-squares-fitting, yielding $C_1 = 1.2\,\mathrm{m/s}$ and $C_2 = 104\,\mathrm{m/sT^2}$. The case without magnetic field, Δp_0, is characterized by C_1 only so that $\Delta p_0 = \rho_\mathrm{f} v_0 C_1$ can also be related to Darcy's law [92]. This is done in a similar fashion as in Chapter 4. Again, Darcy's law yields

$$v_0 = \frac{1}{C_1}\frac{\Delta p_0}{\rho_\mathrm{f}} = k_f \frac{\Delta p_0}{\rho_\mathrm{f} g H_\mathrm{packing}},$$ (7.4)

where $H_\mathrm{packing} \approx 6D = 0.012\,\mathrm{m}$ is the thickness of the packing. Comparing equation (7.3) with (7.4) for $\Delta p = \Delta p_0$ yields $k_f = g H_\mathrm{packing}/C_1$. For the simulation conducted, this results in a value of $k_f \approx 0.1\,\mathrm{m/s}$. Independently, k_f can also be determined using a relation for randomly packed particles [92] and a viscosity correction according to [90] giving $k_f = 0.09\,\mathrm{m/s}$. The value obtained from the simulation is close to this empirical one, showing similar to Chapter 4 that indeed the simulation is realistic.

The agglomerated bubbles are now characterized in terms of their mobility and density. The mobility is quantified by the standard deviation of the velocity of the bubbles when they oscillate around their mean position. This quantity, reported in Figure 7.5, is obtained by averaging in time and over all mobile bubbles. Each symbol in Figure 7.5, 7.6 and 7.7 corresponds to one simulation in Figure 7.3.

In Figure 7.5, an increase of mobility with increasing pressure drop is clearly visible. Furthermore, the magnitude of the fluctuations in both horizontal directions,

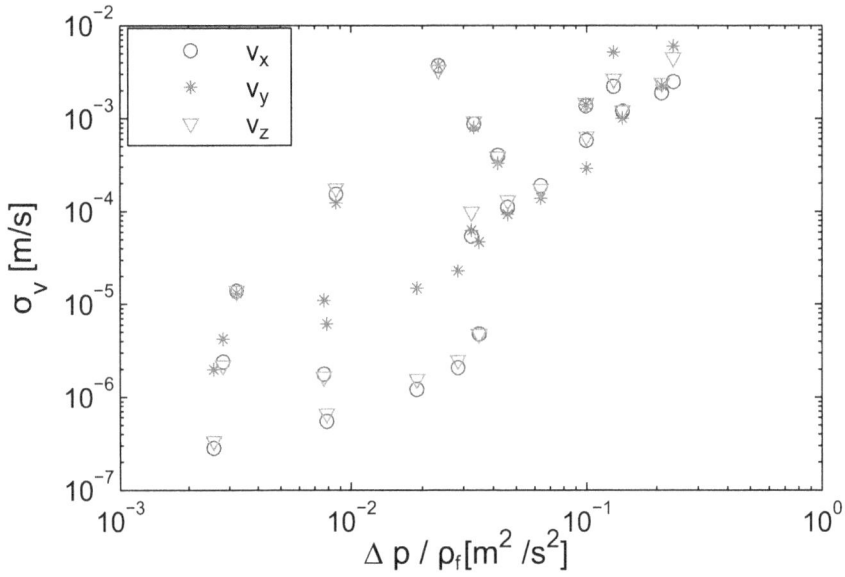

Figure 7.5: Standard deviation of the three velocity components versus pressure drop.

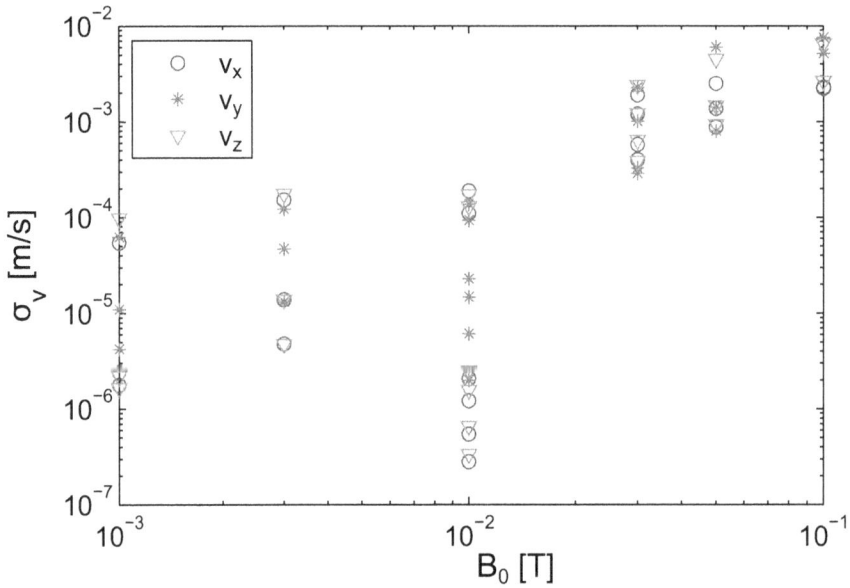

Figure 7.6: Standard deviation of the three velocity components versus strength of magnetic field.

x and z, is very similar. This isotropy is remarkable since the magnetic field is oriented in z-direction. It should hence damp fluctuations in x-direction more than in z-direction as observed for wakes [56]. In most cases the mobility in vertical direction (y) is higher, which is expected as the fluid drag force dominantly acts in this direction. An increase of the mobility with increasing strength of magnetic field has also been found as reported in Figure 7.6. This fact is in contrast to the general observation that magnetic fields damp movement of conducting fluids [22, 23, 25] and might arise from the fact, that the magnetic field increases the pressure drop and thus the mobility according to Figure 7.5.

Averaging the gas fraction in time and horizontal directions and plotting against the vertical direction it is found that the packing density of the mobile bubbles is practically constant over height so that subsequently the gas fraction Φ is determined by averaging the position of the lowest five mobile bubbles to get the volume occupied by the mobile bubbles. Figure 7.7 shows that Φ is almost independent of the pressure drop over a wide range of local pressure drops ($\Phi \approx 0.62$ for $\Delta p/\rho_f \approx 10^{-2}\,\mathrm{m^2/s^2}$). For medium pressure drop the gas fraction decreases because of increasing bubble motion.

Against this general trend, three simulations lead to higher gas fraction, marked in

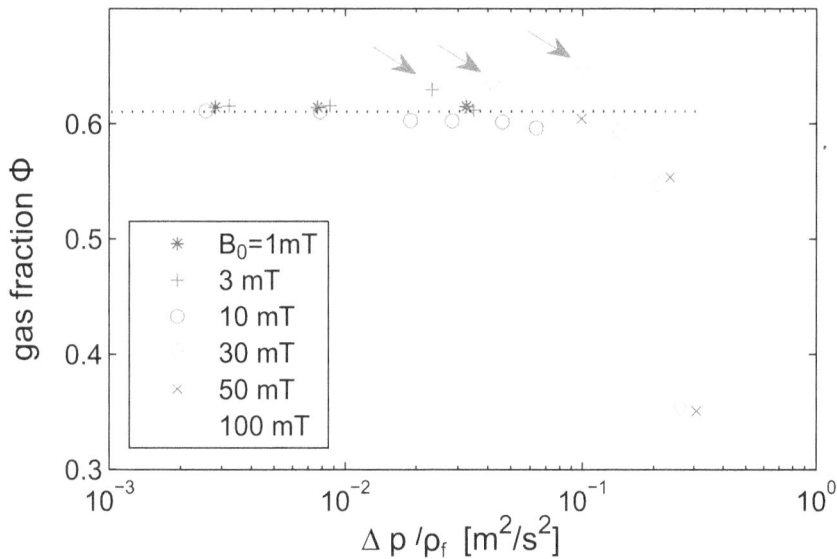

Figure 7.7: Volume fraction of mobile bubbles versus pressure drop. The dotted line indicates the gas fraction at lowest pressure loss. Arrows point to cases with particularly high crystalline order.

Figure 7.7 with an arrow. Analysing local neighbourhood conditions it is found that this reflects an arrangement in crystalline structure to a larger extend than in the other cases, as presented in Figure 7.8. No systematic dependency of this feature

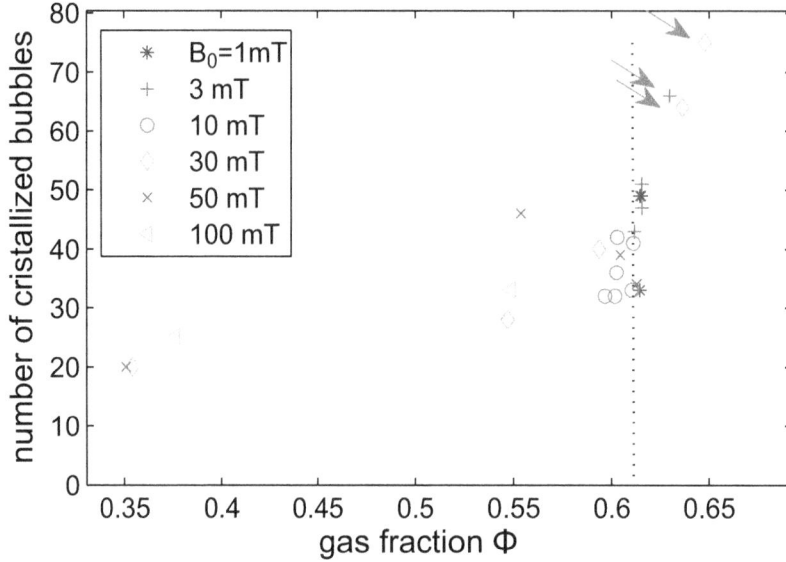

Figure 7.8: Number of bubbles in crystalline arrangement versus volume fraction of mobile bubbles. The dotted line indicates the gas fraction at lowest pressure loss. Arrows point to cases with particularly high volume fraction.

on parameters of the simulation is found. Higher pressure drop creates higher forces on the bubbles and makes the bubbles float which, in turn, strongly decreases the gas fraction.

7.4   Conclusions

In the regime of agglomerated bubbles (see also Chapter 4), the pressure drop depends linearly on the drainage velocity. The results are in good agreement with Darcy's law which, again, validates the approach. A relatively low horizontal magnetic field can increase the pressure drop significantly which furthermore is shown to be proportional to the square of the strength of the field. The quadratic dependency is expected, as presented in the motivation of this chapter. It can be derived from the relations described in Section 7.1 without external electric field E_0:

$$\Delta p - \Delta p_0 \propto f_{\mathrm{L}} \propto B_0\, j \propto B_0(E_0 + v_0\, B_0) \propto v_0\, B_0^2 \tag{7.5}$$

The actual pressure drop $\Delta p - \Delta p_0$ can be compared to the pressure drop $\Delta \tilde{p}$ that would result from downward plug-flow through the domain with velocity v_0

$$\frac{\Delta \tilde{p}}{\rho_{\mathrm{f}}} = \frac{f_{\mathrm{L}} L_y}{\rho_{\mathrm{f}}} = \frac{\sigma_{\mathrm{el}} L_y v_0 B_0^2}{\rho_{\mathrm{f}}} = 5.3\,\mathrm{m/sT^2} v_0 B_0^2 \tag{7.6}$$

$$\frac{\Delta p - \Delta p_0}{\rho_{\mathrm{f}}} = C_2 v_0 B_0^2 = 104\,\mathrm{m/sT^2} v_0 B_0^2. \tag{7.7}$$

Obviously, the actual pressure drop is by a factor of 20 larger than it would be in case of plug-flow. The reason is, that in drainage through the bubble cluster much higher velocities occur due to the small diameter of the flow channels within the cluster. Therefore, the magnetic field can very efficiently increase the corresponding pressure drop. It is to mention, that a purely magnetic field cannot completely stop the drainage, since the pressure drop is also proportional to the fluid velocity. If there is no drainage flow, there would be no force that counteracts gravitation. Also, a purely magnetic field does not allow for a good control of the bubble distribution or even an artificial shifting of bubbles.

The amount of bubbles in crystalline arrangement depends on the pressure drop and the magnetic field strength. In some cases, when the pressure drop is relatively high, but still not high enough to cause flotation of the bubbles, an increase in the number of crystallized bubbles is found. This medium pressure drop corresponds to low drainage velocity and relatively high magnetic field strengths. However, since the appearance of this behaviour is very random and also relies on the drainage flow, it is most likely not applicable for industrial foam generation.

In conclusion, the horizontal magnetic field might be an interesting opportunity for slowing down drainage, but it does not completely fulfil the requirements formulated above. Therefore, in the next chapter, the combination of external electric and magnetic fields will be investigated.

8 Manipulation of foam with electromagnetic fields

8.1 Motivation

As demonstrated in Chapter 7, a purely magnetic field can decrease, but not completely overcome drainage. When the drainage velocity vanishes, it cannot impose a Lorentz force that counteracts gravitation. The purely magnetic field is also not very effective or reliable in manipulating the agglomeration structure of the generated foam.

The next logical step is to superpose a magnetic and an electric field. The electric field causes an electric current. The combination of electric current and magnetic field generates a Lorentz force, that can be used to counteract gravitation and to distribute the bubbles in a desired way. In contrast to a purely magnetic field this mechanism can cause forces even if the velocity of the fluid vanishes. Therefore, it is qualified to completely stop drainage. But due to magnetic induction the flow of the liquid metal generates additional electric currents, causing complex interaction mechanisms which will be investigated in this chapter.

8.2 Physical background

Let us consider a container, filled with liquid aluminium and assume a constant current $\mathbf{j} = j_0\mathbf{e}_x$ and a constant magnetic field $\mathbf{B} = -B_0\mathbf{e}_z$. The resulting Lorentz force

$$\mathbf{f}_{L,0} = -j_0 B_0(\mathbf{e}_x \times \mathbf{e}_z) \tag{8.1}$$

can then be used to counteract the gravitational acceleration $\mathbf{g}$. Indeed, if

$$\frac{\mathbf{f}_{L,0}}{\rho_f} = g_L\mathbf{e}_y = -\mathbf{g} \tag{8.2}$$

with ρ_f the fluid density and $g_L = j_0 B_0/\rho_f$ the Lorentz acceleration, the liquid does not experience any resulting volume force. Without that force, a bubble inside the liquid would not rise but remain floating.

Unfortunately, the presence of bubbles alters the electric field, since their electrical conductivity is much lower than that of liquid aluminium. As a consequence, the electric current is diverted around the bubbles and is no longer constant in space. The magnetic field, in contrast, remains practically unchanged, since the magnetic susceptibility of both the gas and the liquid aluminium is close to 0. The magnetic field can thus be assumed constant in space and time. This situation is depicted in Figure 8.1(a).

For a theoretical investigation, let us consider a single, perfect insulating bubble in an infinite container with horizontal coordinates x and z and vertical coordinate y (Figure 8.1). According to the above argument, the magnetic field is assumed to be constant, $\mathbf{B} = -B_0 \mathbf{e}_z$. With an electric current in the positive x-direction, the Lorentz force is directed upwards. Assuming a perfectly insulating bubble, an analytical expression for $\mathbf{j}$ can be provided superimposing an electric dipole and a constant field $j_0 \mathbf{e}_x$, yielding [27]

$$\mathbf{j} = j_0 \mathbf{e}_x - \frac{j_0 D^3}{16} \left(3 \frac{\mathbf{e}_x \cdot \mathbf{r}}{r^5} \mathbf{r} - \frac{1}{r^3} \mathbf{e}_x \right), \tag{8.3}$$

where $\mathbf{r} = x\mathbf{e}_x + y\mathbf{e}_y + z\mathbf{e}_z$ is the vector pointing from the centre of the bubble to the point considered, r its absolute value, and D the diameter of the sphere. With the constant magnetic field, the Lorentz force is

$$\mathbf{f}_{\mathrm{L}} = j_0 B_0 \left[\mathbf{e}_y (1 + \frac{D^3}{16 r^3}) + \frac{3}{16} \frac{D^3}{r^5} (\mathbf{e}_x \cdot \mathbf{r})(\mathbf{r} \times \mathbf{e}_z) \right]. \tag{8.4}$$

The curl of $\mathbf{f}_{\mathrm{L}}$ can be derived, using the Grassmann identity and the product rule

$$\nabla \times \mathbf{f}_{\mathrm{L}} = \nabla \times (\underline{\mathbf{j}} \times \mathbf{B}) = (\mathbf{B} \cdot \nabla) \cdot \underline{\mathbf{j}} + (\nabla \cdot \underline{\mathbf{B}}) \cdot \mathbf{j} - (\mathbf{j} \cdot \nabla) \cdot \underline{\mathbf{B}} - (\nabla \cdot \underline{\mathbf{j}}) \cdot \mathbf{B}. \tag{8.5}$$

In this notation, the derivation ∇ acts on the underlined vector. The second term on the right-hand side equals zero, because there are no magnetic monopoles ($\nabla \cdot \mathbf{B} = 0$). The third term equals zero, because spatial derivations of the homogeneous magnetic field vanish. The fourth term also equals zero, because due to charge conservation the electric current is free of divergence ($\nabla \cdot \mathbf{j} = 0$). The first term on the right-hand side does not necessarily vanish. For the Lorentz force given in Equation (8.4) it equals

$$\nabla \times \mathbf{f}_{\mathrm{L}} = \frac{3 j_0 B_0 D^3}{16} \begin{pmatrix} \frac{-5zx^2}{r^7} \\ \frac{-5zxy}{r^7} \\ \frac{1}{r^5} \left(x + z - \frac{-5 z z^2}{r^2} \right) \end{pmatrix}. \tag{8.6}$$

This value does not equal 0 in general. Therefore, the Lorentz force is not irrotational and can not be compensated by a pure pressure field. This results in a rotating flow in the vicinity of the bubble. Figure 8.1(b) shows the Lorentz force field $\mathbf{f}_{\mathrm{L}}/\rho_{\mathrm{f}}$ that results from the configuration in Figure 8.1(a), computed with PRIME. For better visibility, the upward component $j_0 B_0/\rho_{\mathrm{f}} \mathbf{e}_y$ is substracted from the force field. One can clearly see the curling character of the Lorentz force.

Due to this curling force field and the resulting curling flows, complex interaction between the bubbles and with wall are to be expected. Sellier et al. [133, 134] investigated these mechanisms for basic situations by analytical considerations, assuming

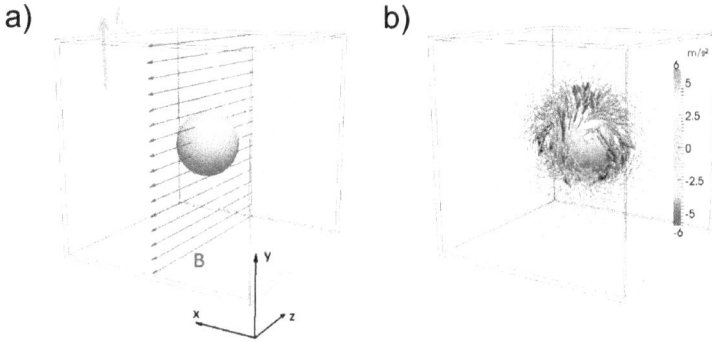

Figure 8.1: Situation of a single stagnant bubble in quiescent fluid. A horizontal electric current j and a perpendicular horizontal magnetic field B result in a Lorentz force f_L, pointing upward. The electric current, however, is diverted by the electrically insulating bubble (a). Lorentz force divided by density, $\mathbf{f}_L/\rho_f$, resulting from the configuration in the left picture. The homogeneous vertical fraction $j_0 B_0/\rho_f \mathbf{e}_y$ has been subtracted to reveal the rotational contribution. The colour corresponds to the vertical component of $\mathbf{f}_L/\rho_f$ (b).

vanishing Reynolds numbers and vanishing electric induction. Here, however, we aim at foam generation so that large numbers of bubbles with finite Reynolds numbers have to be considered, which can no longer be handled analytically. To investigate this situation, phase-resolving simulations with up to 100 bubbles are conducted.

8.3 Setup

In this chapter, three kinds of setup are considered. Setup 1 employs a relatively small domain, aiming to investigate the behaviour of a single bubble. The domain of Setup 2 is bigger, in order to investigate the interaction of two bubbles without being influenced by the walls. Setup 1 has been used in Section 8.4 and Setup 2 in Section 8.5 of this chapter. Setup 3 employs an even bigger domain, designed to investigate the complex interaction of 100 mobile bubbles, representing the situation in realistic foam generation. Setup 3 has been used in Section 8.6 of this chapter. Apart from the geometry, all three setups apply the same physical and numerical parameters, given in Table 8.1.

All three domains are bounded by walls in each direction. The walls perpendicular to the external electric field are perfectly electrically conducting, the other walls are electrically insulating. For the fluid a no-slip boundary condition is set at all walls. Since in this study the collision of bubbles plays a minor role, the basic collision model is employed, as described in Section 3.2. In order to resolve the flow and electric currents in the small channels between the bubbles, a very high numerical resolution is used.

Table 8.1: Physical and numerical parameters of the computational setup.

Quantity	Symbol	Value
Domain size Setup 1	$L_x \times L_y \times L_z$	$3.2D \times 3.2D \times 3.2D$
Domain size Setup 2	$L_x \times L_y \times L_z$	$6.4D \times 6.4D \times 6.4D$
Domain size Setup 3	$L_x \times L_y \times L_z$	$6.4D \times 12.8D \times 6.4D$
Bubble diameter [m]	D	2×10^{-3}
Number of Bubbles Setup 1	N_{b}	1
Number of Bubbles Setup 2	N_{b}	2
Number of Bubbles Setup 3	N_{b}	100
Material properties	Aluminium	Table 1.1
Spatial resolution [m]	Δs	5×10^{-5}
Time step [s]	Δt	5×10^{-5}
Lagrange points	n_{L}	5024
Magnetic field strength [T]	B_0	10^{-1}
Electric current density [A/m²]	j_0	$0 \dots 4.8 \times 10^5$

8.4 Behaviour of a single bubble

To understand the complex interactions between a large number of bubbles it is reasonable to start with the investigation of the behaviour of a single bubble. It is placed in the centre of a domain and is exposed to the electromagnetic field described above. The setup is shown in Figure 8.2. Depending on the Lorentz acceleration

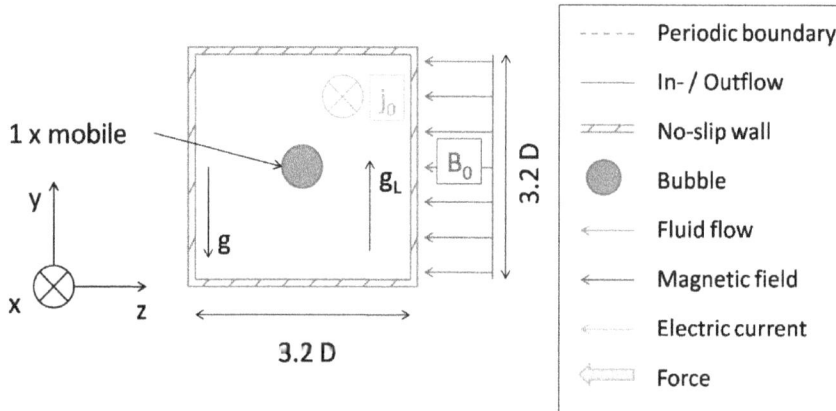

Figure 8.2: Numerical setup (Setup 1) for the investigation of the influence of an electromagnetic field on the trajectory of a single bubble.

g_{L}, the bubble ascends or descends. The vertical velocity of the bubble over time is plotted in Figure 8.3 for different Lorentz accelerations. The initial acceleration of the bubble, extracted from the first 2 ms after initialising, is shown in Figure 8.4. A linear dependence between g_{L} and the resulting initial acceleration of the bubble is observed. Since the latter is a linear function of the electromagnetic force on the bubble, this result is in agreement with Leenov and Kolin [27], proposing a linear dependency between the electromagnetic field strength $j_0 B_0$ and the force on an insulating particle. Figure 8.4 also shows, that perfect flotation is achieved for $g_L = 4/3g$, which corresponds to an overcompensation of gravity, according to Equation (8.2). However, this result also is in agreement with the analytical findings of Leenov and Kolin [27]. Figure 8.3 also shows a non-uniform acceleration

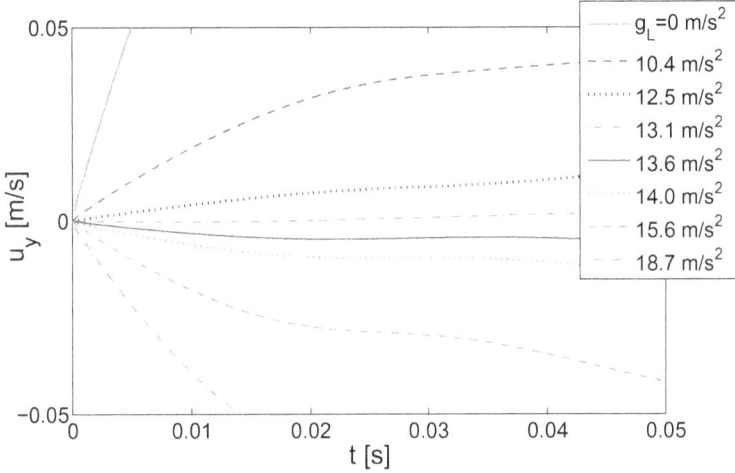

Figure 8.3: Vertical velocity of a single bubble versus time when released under the influence of the considered electromagnetic field. Depending on the magnitudes of the fields, measured by $g_L = j_0 B_0 / \rho_f$, the bubble ascends, floats or descends.

of the bubble. Especially, after 0.03 s the slope of the curves, referring to higher Lorentz accelerations, changes substantially. This feature clearly demonstrates the necessity of taking the motion of the surrounding fluid into account. The rotational Lorentz force accelerates the fluid in the vicinity of the bubble as discussed above, resulting in the flow pattern shown in Figure 8.5. Due to the inertia of the fluid, it takes a while until the curling flow around the bubble is sufficiently fast to influence the movement of the bubble significantly. The curling flow around the bubble has a special shape that can be explained by the modification of the electric current due to the insulation of the bubble. Figure 8.6 shows the x-component of the electric current around a single, insulating sphere. The current is similar to the potential flow around a solid sphere. In positive and negative x-direction there exists a stagnation point (label B) corresponding to zero electric current. In z-direction (actually also in y-direction), there exists a region (label A) where the current density is increased. Therefore, in region A the Lorentz force is stronger than in the rest of the domain, resulting in an upward flow. In region B the Lorentz force is smaller than in the rest of the domain, resulting in a downward flow. Due to the conservation of fluid mass, the streamlines will form closed loops, resulting in the flow structure depicted in Figure 8.5.

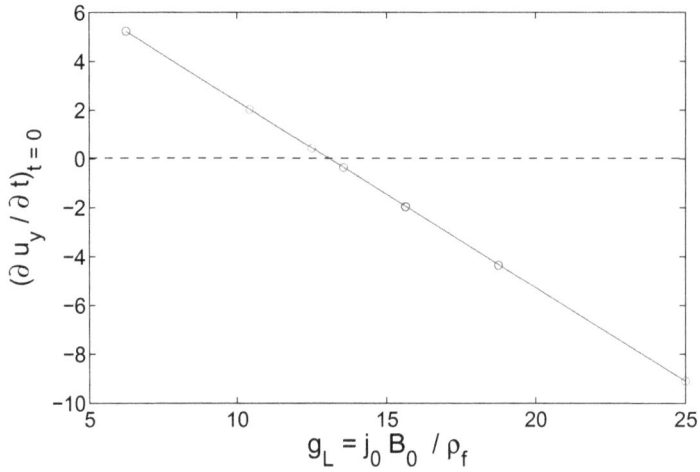

Figure 8.4: Initial acceleration of a single bubble in a horizontal electric current j_0 and a perpendicular horizontal magnetic field B_0. The acceleration varies linearly with g_L and vanishes for $g_L = 4/3\,g$.

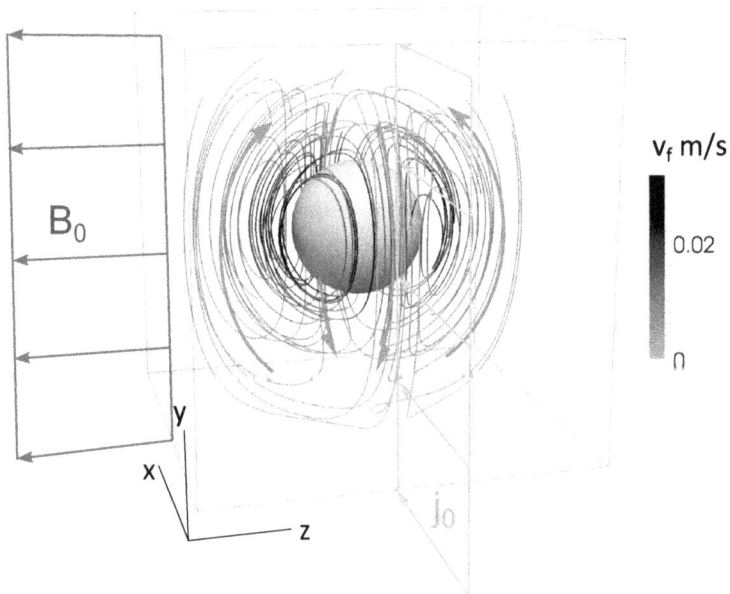

Figure 8.5: Flow field around a stagnant bubble, induced by a horizontal current j_0 and a perpendicular, horizontal magnetic field B_0. Red arrows mark the basic flow structure.

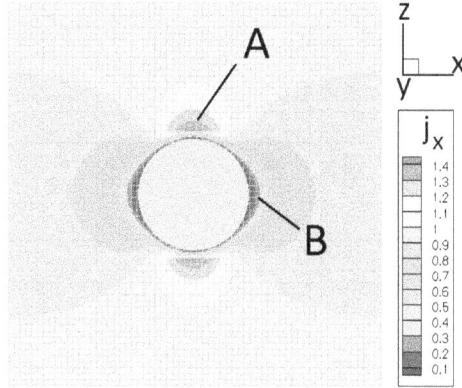

Figure 8.6: Electric current (x-component) around a single, insulating sphere. The electric current density is increased in region A and decreased in region B.

If the bubble is not situated in the centre of the domain but near a wall, it interacts with this wall. Therefore, the fluid flow as well as the electrical field are both altered. To analyse this behaviour, the simulation has been repeatedly initialized with a single bubble, varying the initial position between each run. Figure 8.7 shows the result of 125 runs, plotting the initial acceleration at different initial positions.

While in the centre of the domain the bubble floats, it ascends close to a wall perpendicular to the z-direction with an initial acceleration of $0.005\,\mathrm{m/s^2}$ and descends close to a wall perpendicular to the x-direction with an initial acceleration of $0.0015\,\mathrm{m/s^2}$. This behaviour reflects the impact of the vortex structure around the bubble described above.

Basically, two effects could be responsible for this behaviour. On the one hand, the fluid field could be altered by the wall. On the other hand, the electric current could be altered. If the bubble is placed close to a wall, the no-slip condition at the wall decreases the flow velocity in the gap. Let us assume the bubble to be close to a wall perpendicular to the z-direction. As one can see in Figure 8.5, the flow between bubble and wall would point upward. Therefore, due to friction at the bubble surface this flow helps to accelerate the bubble upward. If this flow is reduced by the friction at the wall, the bubble, initially floating, would move downward. But, as one can see in Figure 8.7, the movement of the bubble shows the opposite trend. A bubble close to a wall perpendicular in z-direction moves upward. Therefore, the modification of the flow field around the bubble can not be at the origin of the bubble-wall interaction. It has to be the modification of the electric current. If the bubble is close to a wall perpendicular to the z-direction, the x-component of the electric current density in region A (see Figure 8.6) is further amplified, because the current is concentrated between wall and bubble. Therefore, the Lorentz force in this region is even higher and thus, the bubble, initially floating, rises. On the contrary, if the bubble is close to a wall perpendicular to the x-direction, the current in region B (see Figure 8.6) is further decreased, resulting in an even lower Lorentz

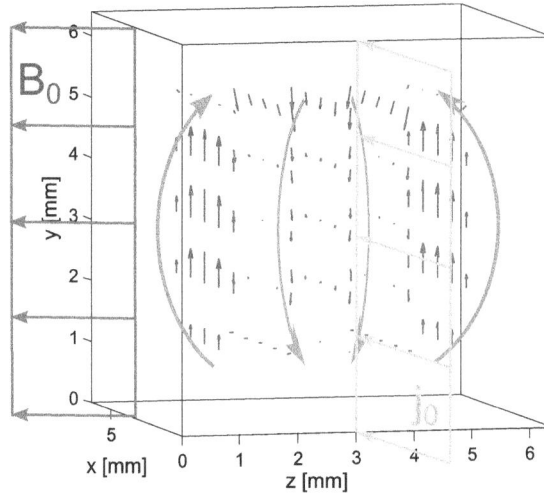

Figure 8.7: Movement of a single bubble released at different positions in a wall-bounded domain. Blue arrows mark the initial acceleration at the referring position, red arrows the overall structure of the acceleration. For $g_L = 4/3\,g$ the initial acceleration equals zero in the centre, but close to a wall the bubble ascends or descends.

force in this region and the bubble, initially floating, would descent. Figure 8.7 also shows, that the bubble-wall interaction only plays a role, if the bubble is very close to a wall. The reason is, that the regions A and B are also very small, decreasing with the square of the distance, as a dipole field does.

8.5 Interaction between a pair of bubbles

Similar investigations are carried out for the interaction of two bubbles. The larger domain of Setup 2 is used now, in order to make sure, that the influence of the walls on the bubbles is negligible. The simulation has been initialized with a single pair of bubbles, such that the centre of gravity of the bubble pair coincides with the centre of the domain. The distance between the bubbles is set to $1.25\,D$ so that the distance of the bubbles surfaces is $0.25\,D$ and the interaction is sizeable. The numerical setup is shown in Figure 8.8. The orientation of the pair varied between each run. The initial acceleration of both bubbles is measured. Figure 8.9 presents the initial acceleration of both bubbles obtained from 42 runs with different orientations of the pair.

In total, 84 arrows are visible, since in each run both bubbles are documented. The two arrows that correspond to the two bubbles in the same run can be found opposing each other. It can be observed that the two bubbles neither attract nor repel each other, but that they accelerate in the same direction. This behaviour

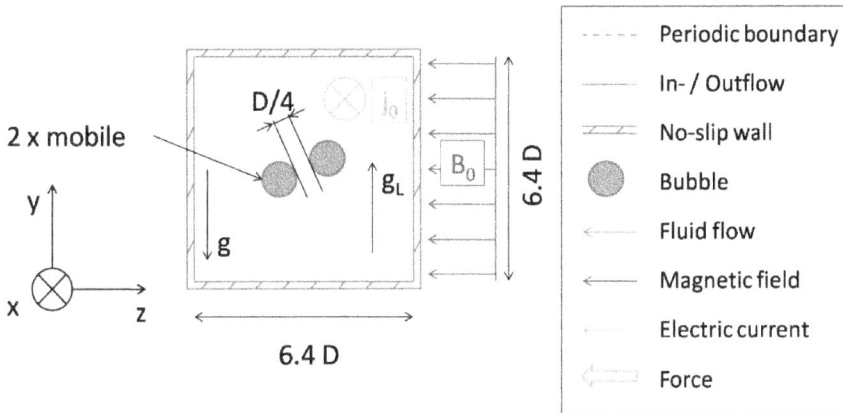

Figure 8.8: Numerical setup (Setup 2) for the investigation of the interaction between two bubbles exposed to an electromagnetic field.

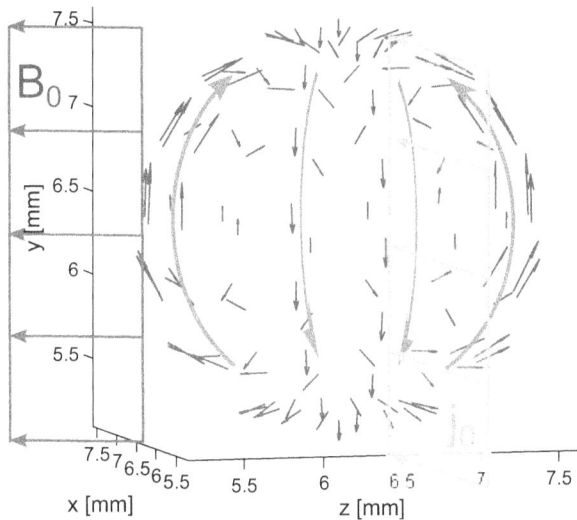

Figure 8.9: Initial acceleration of a pair of bubbles, placed symmetrically with respect to the centre of the domain and released simultaneously. Only the central part of the domain is shown. Blue arrows mark the acceleration at the respective position, red arrows the overall structure of the acceleration.

reflects again the interaction of the electric current around each single bubble. If the distance vector between the centres of the bubbles $\mathbf{x}_{c,2} - \mathbf{x}_{c,1}$ is oriented in z-direction, the electric current between the bubbles, corresponding to region A (see Figure 8.6), is further amplified, resulting in a stronger Lorentz force in this region, pointing upward. Thus, both bubbles, initially floating, are accelerated upwards. If $\mathbf{x}_{c,2} - \mathbf{x}_{c,1}$ is oriented in x-direction, the electric current between them, corresponding to region B (see Figure 8.6), is further decreased, resulting in a smaller Lorentz force in this region and the bubbles move downward.

8.6 Manipulation of bubble clusters

After testing the behaviour of bubbles in simple configurations, the full capacity of the numerical approach is used now to compute the complex interaction inside a large bubble cluster, as it might occur in a technical realization. The setup is shown in Figure 8.10. The parameters are provided in Table 8.1. Please note, that

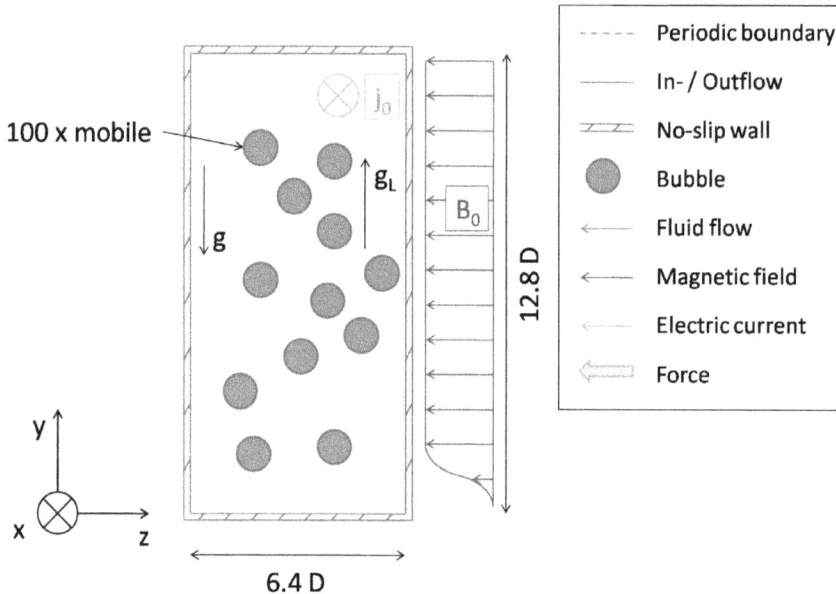

Figure 8.10: Numerical setup (Setup 3) for the investigation of the behaviour of a cluster of 100 bubbles exposed to an electromagnetic field.

the magnetic field is not entirely homogeneous, but smoothly tends to zero in the bottom region as indicated in Figure 8.10. By this decrease, it is possible to realize a lower spatial bound for the bubbles. At five virtual nozzles, located in the bottom of the container, 100 bubbles are released with a total rate of $100\,^1/_s$. They rise and concentrate in the upper part of the domain. When a steady state is reached, the behaviour of the bubbles is averaged and analysed. A snapshot of a steady state for $g_L = 4/3\,g$ is shown in Figure 8.11. Obviously, an upside-down pressure gradient is created by the electromagnetic field. Figure 8.12 shows the average

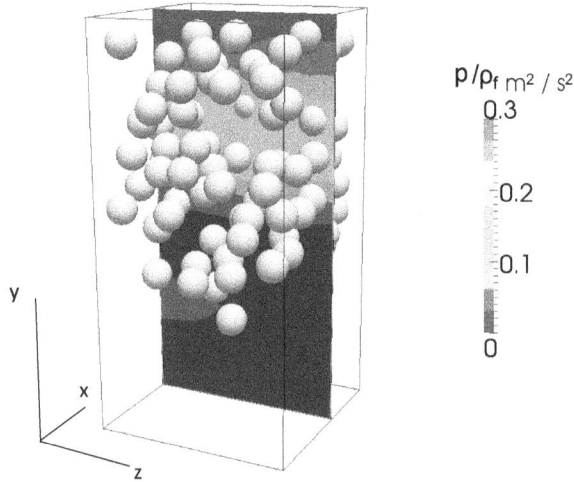

Figure 8.11: Snapshot of 100 bubbles, floating in an electromagnetic field $g_L = j_0 B_0 / \rho_f = 4/3\,g$ in Setup 3. Colour represents p/ρ_f. The gravitational pressure is subtracted in order to reveal the upside-down pressure gradient caused by the electromagnetic field in the upper part of the container that lets the bubbles float.

distribution of the mobile bubbles over the height of the domain for different values of the Lorentz acceleration g_L. At lower electromagnetic field strength, the buoyancy force is not fully compensated. The bubbles hence rise and agglomerate at the ceiling. Nevertheless, close-packed layers, which tend to form in this regime [94], are increasingly less likely to occur with increasing g_L. For $g_L \approx 4/3\,g$ the bubbles start to float and fill the whole upper part of the domain more or less homogeneously. This critical value is in good agreement with the threshold for a floating single bubble presented above. For even higher values of g_L, the bubbles descend, but the region of decreasing magnetic field forms a barrier, since the compensation of gravitation decreases so that the bubbles agglomerate above this region. This barrier appears much softer than a container wall. Thus, no formation of regular packing occurs in this lower part.

The case of flotation is interesting because the bubbles do not rest, but keep on moving. This is due to the complex interaction of the electric current in the vicinity of a bubble and a vertical wall, such that bubbles close to a wall do not float. Instead, bubbles descend close to a wall perpendicular to the x-direction and ascend in the vicinity of a wall perpendicular to the z-direction, creating a global vortex flow in the container, similar to a convection cell. This feature is visualized by the average bubble velocity in Figure 8.13. The mobility of the bubbles is enhanced by the local flow induced in the vicinity of each bubble by the residual Lorentz force depicted in

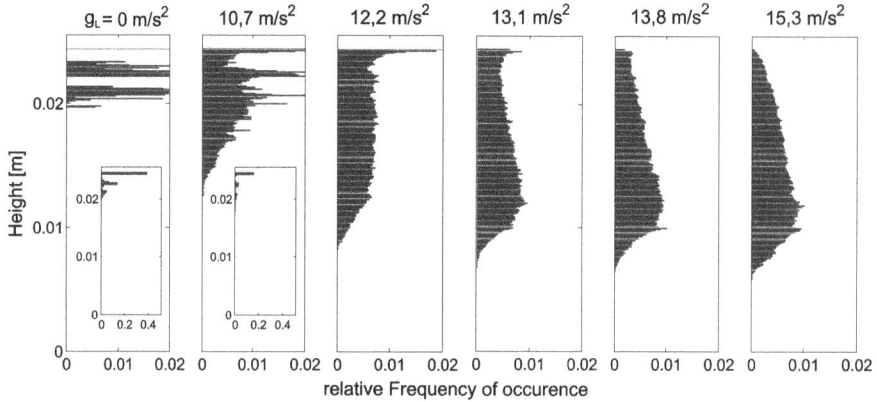

Figure 8.12: Average distribution of 100 bubbles under the influence of the electromagnetic field for different values of $g_L = j_0 B_0 / \rho_f$. As the magnitude of the field approaches 0, layers are formed. Insets with wider horizontal range highlight this feature. For larger magnitudes of g_L bubbles are driven to the bottom of the domain. The decreasing magnetic field in the lower region creates a lower barrier for the bubbles.

Figure 8.1, as well as by the fact that bubbles do not attract each other and thus do not form dense clusters.

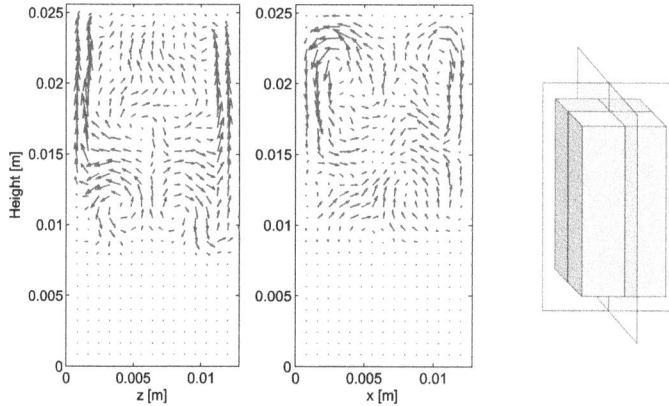

Figure 8.13: Average bubble motion determined by long-time averaging. The vector plots are displayed for a x-y−plane and a x-z−plane both cutting through the centre of the container.

It might be conjectured that the large-scale motion observed would lead to a separation of bubbles with different diameters. This is tested with a simulation identical to the ones presented, apart from replacing the monodisperse bubble swarm by a mix of bubbles with diameters $D = 1.8\,\text{mm}$ and $D = 2.4\,\text{mm}$ for electromagnetic fields generating $g_L = 4/3\,g$. The distribution of both types of bubbles over the height of the domain is shown in Figure 8.14. Obviously, the spatial distribution is equal with small differences only at the top wall and in the decay region of **B** at the bottom. Hence, no vertical separation is observed. The distribution of bubbles is also uniform in any given x-z plane of the domain.

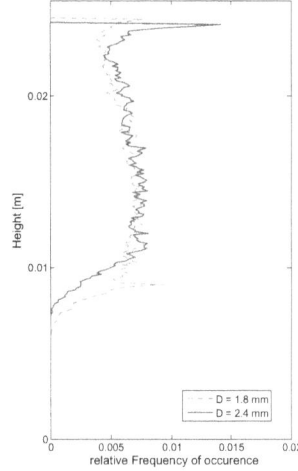

Figure 8.14: Distribution of bubbles in a simulation with two different bubble diameters.

8.7 Artificial bubble distribution

With a homogeneous electric and magnetic field it is possible to distribute the bubbles equally by compensating the homogeneous gravitational acceleration. This raises the question, whether it is possible to apply inhomogeneous fields in order to generate any desired bubble distribution, e.g. concentrate the bubbles in the centre of the domain. This would require a positive source term of the Lorentz force field, which means a positive divergence, as sketched in Figure 8.15(a). Positive sources

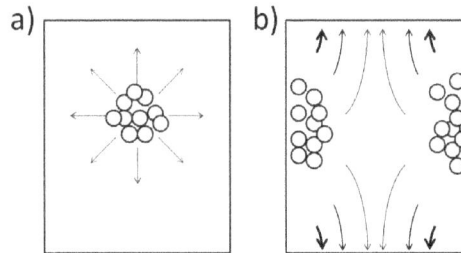

Figure 8.15: Sketch of the force field that causes a certain bubble distribution.

of force generate a local pressure minimum that would attract bubbles, since they are lighter than the surrounding fluid. The Lorentz force is

$$f_L = j \times B. \tag{8.7}$$

Using the commutativity of the triple product and the product rule for derivatives, its divergence $\nabla \cdot \mathbf{f}_L$ can be transformed into

$$\nabla \cdot \mathbf{f}_L \;=\; \nabla \cdot (\mathbf{j} \times \mathbf{B}) \tag{8.8}$$

$$\;=\; \mathbf{B} \cdot (\nabla \times \underline{\mathbf{j}}) - \mathbf{j} \cdot (\nabla \times \underline{\mathbf{B}}). \tag{8.9}$$

Please note that the derivation ∇ acts on the underlined vector. To generate a divergence of the Lorentz force within the container, one needs to apply either a magnetic field or an electric current that is not free of curl.

Both possibilities are now analysed separately. The electric current in a magneto-hydrodynamical flow is given by

$$\mathbf{j} = \sigma_{el}(\mathbf{E} + \mathbf{u} \times \mathbf{B}). \tag{8.10}$$

Its curl is:

$$\nabla \times \mathbf{j} = \sigma_{el}(\nabla \times \underline{\mathbf{E}} + \nabla \times \underline{\mathbf{u} \times \mathbf{B}}) \tag{8.11}$$

Relying on the unsteady flow $\mathbf{u}$ of the fluid in order to control the bubble distribution seems not very promising. Therefore, one would have to rely on the first term on the left-hand side in Equation (8.11). Thus, the curl in the electric current requires curl in the electric field $\mathbf{E}$ and this, according to Maxwell's laws, requires a temporal change of the magnetic field

$$\nabla \times \mathbf{j} \;=\; \sigma_{el}(\nabla \times \underline{\mathbf{E}}) \tag{8.12}$$

$$\;=\; -\sigma_{el}\frac{1}{c}\frac{\partial \mathbf{B}}{\partial t} \tag{8.13}$$

$$\nabla \cdot \mathbf{f}_L \;=\; -\sigma_{el}\frac{1}{c}\mathbf{B} \cdot \frac{\partial \mathbf{B}}{\partial t}. \tag{8.14}$$

Since one would presumably want the divergence in the Lorentz force to remain constant during the whole solidification process of the melt, one would have to apply a constantly growing magnetic field, of sufficient strength, which is extremely inexpedient.

The second possibility, applying a magnetic field which is not free of curl, is also described by Maxwell's laws

$$\nabla \times \mathbf{B} \;=\; \frac{1}{c}\left(4\pi\mathbf{j} + \frac{\partial \mathbf{D}}{\partial t}\right) \tag{8.15}$$

$$\nabla \cdot \mathbf{f}_L \;=\; -\frac{4\pi}{c}\mathbf{j} \cdot \mathbf{j} \tag{8.16}$$

Following the same argumentation as above, applying electrical fields $\mathbf{E}$ that constantly grow in time is inexpedient. But, a strong electric current $\mathbf{j}$ would induce curling magnetic fields in its surrounding that would result in a local divergence of the Lorentz force. However, the divergence is strictly negative, resulting in the local distraction, not attraction of bubbles. Also, due to the high value of the speed of light c, this method would require extremely strong electric currents. To apply a small force gradient of $df/dx = 0.01\rho_f g/D$ one would need about 10^6 A/m² electric current. Last but not least, the Lorentz force is oriented perpendicularly to the path of the electric current. Therefore, the confinement for the bubbles would only be two-dimensional. They would not be repelled from a point but from a line, limiting the degrees of freedom in applying this method.

The presented arguments demonstrate that the artificial distribution of bubbles in the inside of a domain would be problematic. Nevertheless, it is possible to achieve certain types of bubble distributions by applying Lorentz force fields that are free of divergence. Bubbles would then agglomerate at certain regions close to walls. A possible combination of electric current and magnetic field would be

$$\mathbf{j} = \begin{pmatrix} -j_0 \\ 0 \\ 0 \end{pmatrix}, \mathbf{B} = \begin{pmatrix} 0 \\ (z - z_0)\tilde{B} \\ B_0 + (y - y_0)\tilde{B} \end{pmatrix} \Rightarrow f_L = \mathbf{j} \times \mathbf{B} = \begin{pmatrix} 0 \\ j_0 B_0 + (y - y_0)j_0\tilde{B} \\ -(z - z_0)j_0\tilde{B} \end{pmatrix}.$$

$$(8.17)$$

The electric current and the magnetic field are free of divergence and curl. The resulting Lorentz force distribution is sketched in Figure 8.15(b). This type of field is now tested in a numerical simulation with the setup described above. The Lorentz acceleration $g_L = 4/3g$ is chosen to compensate gravity. The centre of unvaried Lorentz force (y_0, z_0) is placed in the centre of the domain. The field variation $\tilde{B}$ is set to equal $0.02B_0/D$. A snapshot of the resulting distribution of 100 mobile bubbles is shown in Figure 8.16. One can see that the bubbles are driven towards the left and right wall. Of course, this shape of electric and magnetic field is only one example. Depending on the application, a favourable bubble distribution can be generated by applying an according combination of electric and magnetic field. But, as shown above, due to the absence of divergence in the force field bubbles cannot be agglomerated within the bulk of the domain but only at walls.

8.8   Conclusions

The influence of an electromagnetic field on simple and complex bubble arrangements has been studied. A horizontal electric current and a perpendicular, horizontal magnetic field have been considered. The combination of these fields creates a Lorentz force that points upward and counteracts the gravitational force. Single bubbles start to float at a magnitude of the fields of $j_0 B_0/(\rho_f g) \approx 4/3$, which is in agreement with analytical studies by Leenov and Kolin [27].

Due to the non-vanishing curl of the resulting Lorentz force field, a complex flow vortex is generated in the vicinity of a bubble. The variation of the electric current results in a distinct interaction scheme for bubble-wall and bubble-bubble interaction. A pair of bubbles subjected to the electromagnetic field neither attract nor repel each other, which is an important point in generating homogeneously distributed bubbles with this method.

Simulated clusters of 100 bubbles float under the same conditions and keep moving due to the finite curl of the induced Lorentz force in the vicinity of the bubbles.

Figure 8.16: Snapshot of the distribution of 100 bubbles floating under the influence of an electromagnetic field given in Equation (8.17).

The interaction with the walls generates a global convection structure. A mix of two types of bubbles with different diameters is not graded according to their size by this convection, which is also important for generating homogeneous bubble distributions in systems with polydisperse bubbles.

Unfortunately, it is not possible to make the bubbles agglomerate somewhere within the bulk of the liquid metal, because it is inexpedient to generate a Lorentz force that offers a positive divergence. The only thing that can be done is to drive the bubbles toward distinct regions of the the wall. It is also possible to separate the cluster of bubbles and drive them towards different wall regions. This contradicts the classical approach in light-weight construction, as it is sketched in Figure 8.17.

In the inner regions of a construction element, usually lower stresses attack. For

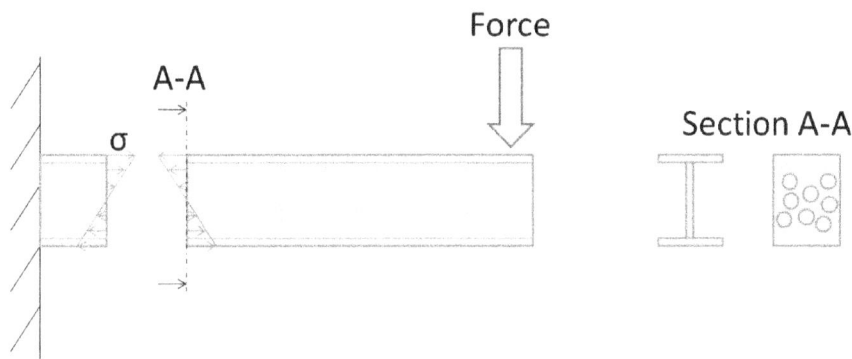

Figure 8.17: Classic approach in light-weight design. In the inside of construction elements usually lower stresses occure (left). Consequently, material is reduced in these regions by double-T beams or sandwich-design metal foam (right).

example, in the middle of a beam with perpendicular tip force the longitudinal stress equals zero. Therefore, material can be reduced in the middle, as it is done with a double-T-profile. Placing bubbles in the middle of a beam would also be advantageous for the specific stability of the element, as it is done in the sandwich-design mentioned in the introduction. But this is not feasible with th method investigated. Placing bubbles in the outer regions with high stresses will weaken the element, which is usually not desired. However, if one tries to design a particularly soft metal device, e.g. to replace bones [135–137], this technique might be relevant.

In this chapter relatively high electromagnetic fields are used, since the aim is to completely compensate gravity. However, one other aspect of this configuration is the influence of lower field strengths on the agglomeration process. In Figure 8.12 one can see that lower field strengths do not completely prevent the bubbles from agglomeration, but they most certainly influence the resulting bubble structure. Maybe the resulting metal foam has particularly high or particularly low amounts of crystalline ordering. This question will be addressed in Chapter 9.

9 Influence on foam structure

9.1 Motivation

In Chapter 8 the influence of a horizontal electric current and a perpendicular, horizontal magnetic field on the agglomeration of bubbles has been reported. The electromagnetic field causes a Lorentz force, pointing upward, that counteracts gravitation. Therefore, the bubbles start to float and distribute equally. If a lower field strength is applied, gravitation is not counteracted completely. Therefore the bubbles will rise and agglomerate at the top of the domain. As shown in Section 8.4, curling Lorentz forces are generated in the vicinity of a bubble. Also the rising velocity and thus, the impact and the time scale of the agglomeration is modified by the electromagnetic field. Consequently, it is possible that the electromagnetic field can modify the amount of crystalline arranged bubbles or even the structure of the crystallisation. Consequently, the mechanical properties of the resulting metal foam would be changed, as it is found in Chapter 6. This will be investigated in this chapter.

9.2 Setup

The focus of this chapter is on crystallisation. Therefore, the setup from Chapter 5 is adopted, featuring one closed-packed layer of 52 fixed bubbles on top, in order to trigger crystallisation. In contrary to Chapter 5, the horizontal directions are not periodic, but vertical no-slip walls are placed. These walls disturb the regular crystal growth and therefore counteract the triggered crystallisation. In combination, one feature is placed that amplifies crystal growth and one that hinders crystal growth. These features compete against each other. Therefore, this setup is supposed to be very sensitive to the influence of electromagnetic fields on the crystallisation process. A downward drainage flow is not introduced.

Throughout the whole domain, 200 mobile bubbles are placed in a random arrangement and released simultaneously. Similar to Chapter 8, a horizontal electric current $j_0\mathbf{e}_x$ and a perpendicular, horizontal magnetic field $-B_0\mathbf{e}_z$ are applied. These fields cause a Lorentz force that points upwards or downwards, and therefore counteracts or amplifies gravitation. In accordance to Chapter 8 the combined field strength is expressed by an acceleration g_L that is caused by the Lorentz force

$$g_L = j_0 B_0 / \rho_f \tag{9.1}$$

If the Lorentz force points upwards and counteracts gravitation, g_L is positive, if it points downwards and amplifies gravitation, g_L is negative. The variation of g_L is carried out by changing the electric current j_0 and keeping the magnetic field B_0 constant. Therefore, even in case of vanishing g_L, a magnetic field is present that impacts on the flow. For each value of g_L, 30 runs with different initial positions of the mobile bubbles have been performed and analysed statistically. A sketch of the setup is shown in Figure 9.1. Physical and numerical parameters are given in Table 9.1.

Table 9.1: *Physical and numerical parameters of the computational setup.*

Quantity	Symbol	Value
Domain size Setup 1	$L_x \times L_y \times L_z$	$7.75D \times 15.5D \times 7.75D$
Bubble diameter [m]	D	2×10^{-3}
Number of fixed Bubbles [-]		52
Number of mobile Bubbles [-]	N_b	200
Material properties	Aluminium	Table 1.1
Spatial resolution [m]	Δs	1.21×10^{-4}
Time step [s]	Δt	1×10^{-4}
Lagrange points	n_L	857
Magnetic field strength [T]	B_0	10^{-1}
Electric current density [A/m²]	j_0	$-2.4 \ldots 2.4 \times 10^5$

In this chapter, the time scale and the mechanics of the collision of bubbles in the agglomeration process has significant influence on the results. Therefore, the advanced collision model of Section 3 is used here. In fact, it has been developed to make this very kind of simulation more realistic and reliable.

9.3   Results

The bubbles rise and agglomerate below the fixed layer. Figure 9.2 shows the vertical position of bubbles over time for different values of g_L. For better visibility, only every third bubble is plotted. Obviously, the average velocity of ascent depends strongly on the electromagnetic field strength g_L. Since the initial bubble distribution is equal for all cases, the amount of bubbles that arrive at the top depends linearly on the rising velocity. Also the kinetic energy and, therefore, the magnitude of the impact of a rising bubble colliding with the existing cluster depends on the rise velocity. One can also see in Figure 9.2 that after the impact of the last bubble, the cluster is not yet stable but certain rearrangements take place. In case of high Lorentz forces, e.g. $g_L = 10 \, \text{m/s}^2$, the cluster will never rest. The curling Lorentz

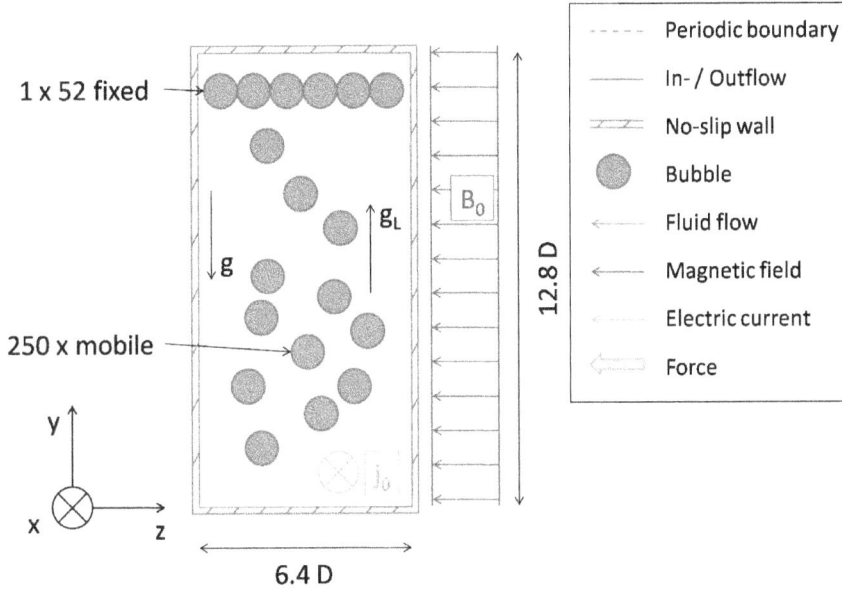

Figure 9.1: Numerical setup for the investigation of the influence of an electromagnetic field on the crystalline arrangement of bubbles.

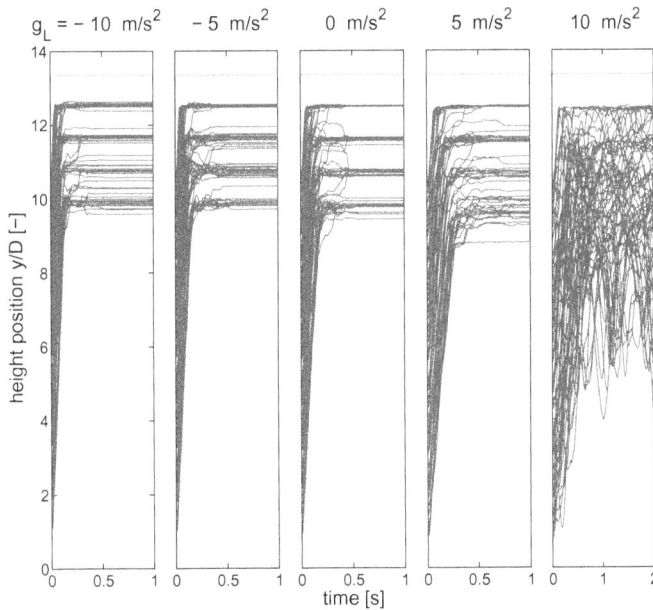

Figure 9.2: Vertical bubble positions over time for different electromagnetic field strength g_L. For visibility, only every third bubble centre is plotted.

forces, described in Section 8.2 keep stirring the cluster. For smaller Lorentz forces, e.g. $g_L = 5\,\mathrm{m/s^2}$, the curling forces result in a longer time period of about $1\,\mathrm{s}$ that the cluster takes for settling. In case of vanishing or negative g_L, settling requires about $0.5\,\mathrm{s}$.

The crystalline arrangement of the resulting cluster is quantified by scanning the theoretical positions of crystalline arrangements for the presence of bubbles. The theoretical positions result from the position of the fixed layer. Due to the softness of the collision model, the vertical distance Δy_c between two layers of bubbles in crystalline arrangement depends on the modified gravitation $g - g_L$ and therefore has to be derived for every value of g_L from the simulation. To do so, the histogram of the height distribution is used, as demonstrated in Figure 9.3. Knowing the theo-

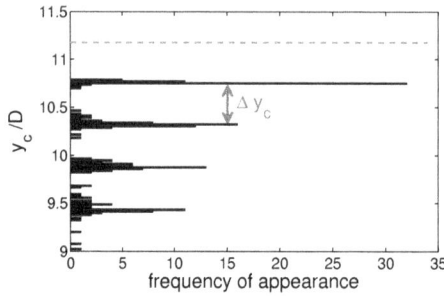

Figure 9.3: Distance Δy_c between the first and the second layer of mobile bubbles, derived from their height distribution. The broken line marks the position of the fixed bubbles.

retical positions, the amount of crystallised bubbles n_{cryst} can be determined. To do so, a distance of $0.1D$ between the centre of the bubble and the theoretical position is tolerated. This results in a probability $P_{\mathrm{false\,positive}}$ for false-positive crystallisation events of approximately

$$P_{\mathrm{false\,positive}} = \frac{V_{\mathrm{crystallised}}}{V_{\mathrm{total}}} = \frac{n_{\mathrm{crystal\,positions}} \cdot V_{\mathrm{tolerated}}}{V_{\mathrm{total}}} = \frac{3 \cdot 4\pi/3 \cdot (0.1D)^3}{\pi/6 \cdot D^3 \cdot 1/\Phi} \approx 0.02. \tag{9.2}$$

The volume $V_{\mathrm{tolerated}}$ in which a bubble is recognised to be crystallised equals $4\pi/3 \cdot (0.1D)^3$. In FCC and HCP, the horizontal position of a layer can correspond to an A, B, or C position as shown in Figure 5.5. This results in $n_{\mathrm{crystal\,positions}} = 3$ positions per bubble space $V_{\mathrm{total}} = \pi/6 \cdot D^3 \cdot 1/\Phi$. The resulting probability equals $2\,\%$ which is negligible. Figure 9.4 shows the ratio n_{cryst}/N_b of crystalline arranged bubbles for different values of g_L, averaged over 30 runs, each. The error bars correspond to the statistical uncertainty of the mean value.

Obviously, the amount of crystallisation becomes smaller with increasing values of g_L. This is surprising. It is expected that with higher g_L the bubbles have more time to settle and to rearrange, get less jammed and therefore more crystallisation should occur. However, the result is the opposite.

Until now, monodisperse bubbles were considered. A slight polydispersity is well known to destroy the crystalline ordering. Since negative values of g_L have been

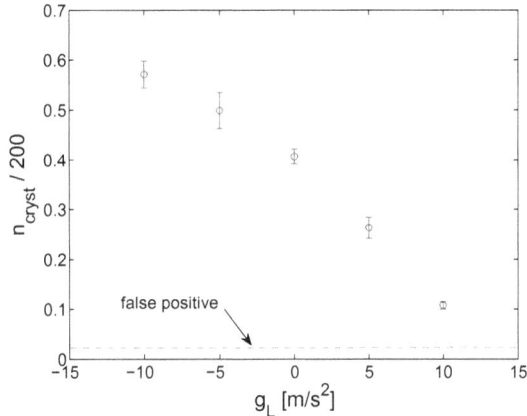

*Figure 9.4: Relative amount of crystalline arranged bubbles, averaged from 30 compu-
tations each. Error bars mark the statistical uncertainty. The broken line shows the
probability for false-positive crystallisation events.*

found to increase crystalline ordering it is interesting to test, if this type of field
could also create crystalline ordering in polydisperse packings. To do so, the setup
above has been modified. The 52 fixed bubbles are still monodisperse, while the
200 mobile ones now have a Gaussian-distributed probability-density function of
the bubble diameter $\tilde{D}$

$$p(\tilde{D}) = \frac{1}{\sigma(D)\sqrt{2\pi}} e^{-0.5\left(\frac{\tilde{D}-D}{\sigma(D)}\right)^2} \tag{9.3}$$

The polydispersity $\sigma(D)$ is set to $0.05D$. The field strengths $g_L = -10\,m/s^2$ have
been applied. The result is, as expected, an unordered packing. The final packing for
monodisperse and polydisperse bubbles at $g_L = -10\,^m/s^2$ is shown in Figure 9.5. At

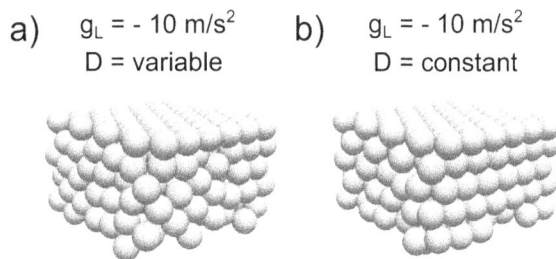

*Figure 9.5: Final arrangement of slightly polydisperse (a) and monodisperse (b) bubbles
under the influence of a downward Lorentz force.*

the same time, in Figure 9.6 the distribution of the centre positions for both cases are
reported by means of a three dimensional scatter plot. One can see that the ordered

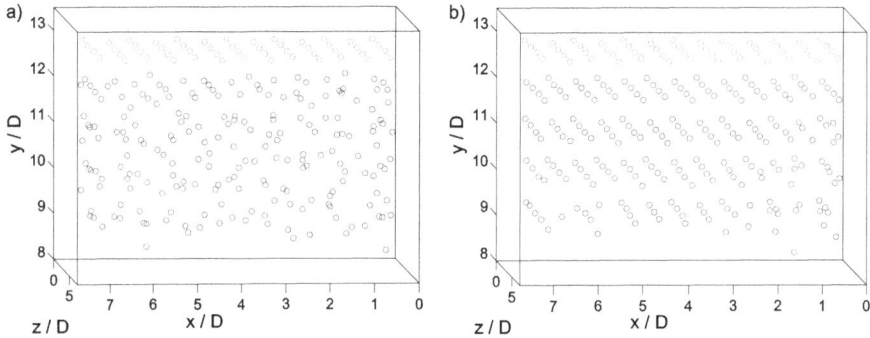

Figure 9.6: Final arrangement of slightly polydisperse (a) and monodisperse (b) bubbles under the influence of a downward Lorentz force.

structure induced by the topmost layer immediately is lost in the polydisperse case. However, due to the variable diameter a quantification of the crystallisation cannot be done in a similar way as above and has been avoided at this point.

9.4   Conclusions

The amount of crystalline ordering can be manipulated by adding electromagnetic fields during the agglomeration process. A field that counteracts gravitation results in a less ordered packing. A field that amplifies gravitation results in a more crystalline ordering. The reason could be that the curling flows tend to mix the bubbles and therefore, destroy crystallisation. Let us consider $g_{\mathrm{L}} = 0\,\mathrm{m/s^2}$ to be the basic case. The bubbles rise and agglomerate. In the middle of the domain, crystalline ordering is dominant, triggered by the layer of fixed bubbles. Close to the side-walls, crystalline ordering is disturbed because bubbles there minimise their local potential energy, sacrificing the global minimum energy state. This state is comparable to a random close-packed state, corresponding to jamming. Adding an electromagnetic field with $g_{\mathrm{L}} < 0\,\mathrm{m/s^2}$ has two effects. Firstly, the gravitation is amplified, resulting in a bigger tendency for the bubbles to minimise energy. Secondly, curling forces are added that could be compared to shaking. Shaking has been reported to increase the amount of crystallisation in the packing of beads [101]. It temporarily overcomes the need of each bubble to minimise only its own energy and therefore, allow the system to rearrange and to enter a global lower energy state, resulting in an increase of crystallisation. On the other hand, in case of positive values of g_{L} the amount of crystallisation is decreased, compared to the basic state with $g_{\mathrm{L}} = 0\,\mathrm{m/s^2}$. Positive values of g_{L} also add stirring flows. However, the need of the system to minimise global energy is reduced at the same time because gravity is counteracted. Therefore, the stirring only prevents the system from entering a low energy state. However, this is only one possible explanation for the numerical result. It is also very qualitatively. Maybe, more complex interaction mechanisms between bubbles, curling flows and walls play an important role. Further investigations are necessary to completely understand this result.

If the size distribution of the bubbles is slightly polydisperse, the crystalline ordering is suppressed. The application of an electromagnetic field cannot restore it. This result is expected, because the unordered packing of polydisperse bubbles does not result from jamming but from geometrical limitations. While the first layer below the fixed one still contains some crystalline arranged bubbles, the regular pattern becomes distorted due to the variation in bubble distances. The next layer forgets about the initially regular arrangement and arranges in a random fashion, resulting in random close-packed structure.

10   Concluding remarks

In this thesis the manipulation of metal foam in its liquid state is analysed numerically. The focus is on the free movement and the agglomeration process of bubbles, forming foam. In this state the foam is very wet, containing more than 25% liquid metal so that the behaviour of the liquid plays an important role. Therefore, phase-resolving simulations are highly relevant to understand the foam behaviour in this state. The transition to dry foam, however, can not be investigated with the present method, because the bubbles are restricted to be spherical. The packing density of 74% marks the most dense sphere packing, which is also the limit for the present method. To simulate higher gas fractions one has to allow eiter for bubble overlapping, which is unphysical, or for bubble deformation. Recently, Schwarz has managed to include deformable bubbles into PRIME [54]. However, these methods are most likely not applicable for the simulation of foam, because they rely on interpolation of a well resolved pressure field in the surrounding fluid, in order to determine the shape of a bubble. This cannot be accomplished in thin lamellas. For the simulation of dry foam, other methods exist, e.g. the Surface Evolver [138–141] or Potts model [142]. These methods do not take fluid dynamics into account but base on Plateau's law [1] and energy minimization.

The fundamental numerical validation methods, such as grid studies and convergence tests have been carried out by the author as well as by the students he supervised and the colleagues at the chair of Fluid Mechanics. The results of these standard practices, however, are not included in this book. Instead, several comparisons with analytical, numerical and experimental data are presented. They are summarized in Table 10.1.

In general, good agreement is found, pointing out a high quality of the numerical

Table 10.1: Numerical results throughout this thesis that allow to validate the numerical approach.

Reference	Setup	Quantities	Section
Experiment	Bubble rebounce at plate	$\mathbf{x}_c(t)$	3.5
Clift et al. [48]	Drag coefficient	c_D	4.2
Darcy's law [90]	Drainage through agglomerated bubbles	$\Delta p\, vs.\, \dot{V}_0$	4.4
Momentum balance	Flotation of bubbles in static drainage	$\Delta p\, vs.\, F_b$	4.5
Several systems [94]	FCC preference in sphere packing	n_{FCC}/n_{HCP}	5.3
Leenov and Kolin	Lorentz force on insulating sphere	$f_L\, vs.\, j_0 B_0$	8.4

data.

In Chapter 3, a collision model is derived, that is easy to implement and yields good agreement to experiments. This collision model might be interesting also for other researchers, investigating the interaction of soft spheres numerically.

In Chapter 4 the movement and agglomeration of bubbles in a downward drainage flow has been investigated. For low drainage velocities, a linear relation between pressure drop and drainage flow has been found. This result is in agreement with Darcy's law [90]. For drainage velocities above a critical value, the bubbles become mobilized and float. In the floating state the pressure drop balances the buoyancy force of the bubbles.

When the bubbles agglomerate, they prefer FCC packing over HCP packing. This has been found by many researchers for different systems, such as packing of glass beads, bubbles or colloids. However, this effect so far could not be explained. In this work, a simple instability mechanism has been identified, that destroys HCP packing more likely and therefore causes FCC to outlive HCP. The question of FCC preference is essentially academic. However, if FCC ordered metal foam has different properties than HCP foam, it might be desired to amplify the generation of FCC or HCP packing. To answer this question, the influence of the bubble structure on the mechanical properties of solid foam have been investigated with a Finite-Element method, implemented in ANSYS FEM. These simulations have been able to identify a small difference in the Young's modulus between FCC and HCP foam. However, this difference is below 6 % which does not justify the effort of controling the structure.

Adding a horizontal magnetic field to a downward drainage flow through agglomerated bubbles significantly increases the drainage resistance. The increase of the pressure drop for a given drainage velocity depends quadratically on the applied magnetic field strength. This quadratic dependency is well known from the Hartmann flow. A magnetic field of 0.1 T was found to double the drainage resistance. A magnetic field on 1 T increases it by a factor of 100. Hence, a horizontal magnetic field is qualified to slow down drainage in wet metal foam. Apart from that, it has only low influence on the bubble agglomeration and the resulting foam structure.

An even more promising approach is the combination of the horizontal magnetic field with a perpendicular, horizontal electric current. This combination causes a Lorentz force that can counteract gravitation. Consequently, the bubbles do not rise but float in the domain distributing almost homogeneously. There are some special features of this state that support the equal distribution. Two bubbles, that are close to each other, do not attract each other. So the bubbles do not agglomerate with each other. A bubble close to a wall is not attracted by it, so there is no agglomeration at the walls, neither. Finally, a curling flow is induced in the vicinity of each bubble, causing a global stirring motion that increases homogeneity of the distribution. The equal distribution of bubbles offers a very interesting opportunity to customize metal foam. Usually, the gas fraction of metal foam is more or less fixed. The bubbles rise, agglomerate and form dense clusters with gas fractions above 0.64 %. Drainage further increases the gas fraction. But in the floating state one can control the gas fraction simply by the amount of gas one adds to the liquid metal. The bubbles distribute homogeneously due to the electromagnetic field. In that way, one can generate porous metal with any gas fraction between 0.0 and 0.64 %. This is very handy for a designing engineer who could choose the stability of the material according to the requirements.

It would be even better, if one could generate not only homogeneous, but any bubble distribution. Knowing the regions of low stress in a construction element, one could arrange bubbles on purpose in these regions, minimizing the weight of the element while meeting the stability requirements. However, due to the properties of electro-magnetic fields, it is not possible to assemble bubbles somewhere in the inside of a container. The only option is to assemble bubbles close to walls. Unfortunately, this usually weakens construction elements significantly. Therefore, the technique of customized distribution most likely is only relevant for rare, specialized elements that ought to be soft or deformable.

Employing FEM simulation it was found that the Young's modulus of FCC or HCP at maximum packing density is by a factor of three smaller than the Young's mod-ulus of RCP at maximum density. To generate particularly soft material, it would be desired to enforce or increase crystalline packing. It was discovered, that an electromagnetic field of $g_L = -10 \, \mathrm{m/s^2}$, which does not counteract but amplify the gravitational acceleration, can increase the amount of crystallization by about 50 percent. The induced curling Lorentz force has an effect similar to shaking, that increases the packing density. The reasonability of this technique for practical appli-cation is very questionable. Amplifying gravity also means to increase the drainage and thus, the foam very soon transforms from wet foam to dry foam. In case of dry foam, the bubbles are no longer spherical and therefore do not prefer HCP or FCC packing, but transform to BCC with Kelvin-cell shaped bubbles [1, 3]. To actually use the technique of increased crystalline ordering, the solidification has to take place very rapidly after introducing the bubbles. On the other hand, a counteract-ing electromagnetic field of $g_L = +5 \, \mathrm{m/s^2}$ does not increase but decrease crystalline ordering. The decrease most likely has no practical relevance as well, because if one wants less crystallization, one simply reduces the effort of generating monodisperse bubbles, which in a very cheap and effective manner prevents crystalline ordering.

Even though electromagnetic fields can positively influence the fabrication of metal foam, it is not so easy to apply these fields. For the magnetic field it would be difficult to apply permanent magnets, because the melting temperature of alu-minium ($660°C$) is well above the Curie temperature of typical ferrite materials ($100 \ldots 460°C$), so they would loose their magnetization. Also, the application of super-conductive coils close to hot liquid aluminium might be difficult. Conse-quently, the application of classic coils would be necessary, consuming substantial electric power. The assumed magnetic field strength of 0.1 T is moderate and can be generated by classic coils also in industrial scales. The power consumption of a coil generating 0.1 T within a volume of $10^{-3} \, \mathrm{m^3}$ would consume about 10 kW [143], which is practicable.

The required electric current for levitation of liquid aluminium can be estimated by

$4/3\rho_f g = j_0 B_0$, yielding $0.32\,\mathrm{A/mm^2}$. For a construction element of $10\,\mathrm{cm} \times 10\,\mathrm{cm} \times 10\,\mathrm{cm}$ one would need $3200\,\mathrm{A}$. This value is quite large, but not untypical for industrial applications. For example, the electric current for smelting aluminium in the Hall-Héroult process can equal about $200\,000\,\mathrm{A}$ [144]. Due to the high electric conductivity of liquid aluminium the corresponding electric power for this construction element would be $200\,\mathrm{W}$, which is quite feasible, at least for small geometries.

Still, open questions remain. This work focuses on wet foam. The influence of electromagnetic fields on dry foam might be different. The principal mechanisms are most likely also applicable to dry foam. A horizontal magnetic field increases the drainage resistance. The combination of magnetic and electric field causes Lorentz forces that compensate gravitation. Especially, the magnetic field is not influenced by the amount of gas, because gas and liquid aluminium both have a magnetic susceptibility close to zero. However, the electric current is very sensitive to the gas fraction [29]. For low fractions of spherical bubbles the electric current can be described more or less to be a superposition of electric dipoles, as suggested by Equation (8.3). However, in dry foam the electric current to a large extend channels along the Plateau borders. This significantly changes the shape of the Lorentz force distribution. The Plateau borders form a network of conducting wires that experiences in a magnetic field an upward Lorentz force. Curling flows hardly appear. Therefore, the quantitative findings of the behaviour of foam in magnetic and electromagnetic fields most certainly differ between wet and dry foam.

Another open question is the influence of temperature and solidification on the foam. One aspect is the strong dependency of the viscosity of liquid aluminium on the temperature [45]. If strong temperature gradients exist, this might unbalance viscous forces and therefore cause additional bubble movement. The influence of the temperature-dependent viscosity on the bubble distribution could be investigated with the present method. Another aspect is the solidification front. It propagates through the foam and maybe attracts or repels bubbles and therefore, manipulates the bubble arrangement. Simulating the solidification front is more challenging, because solidification means a jump in the material properties. Various methods for the simulation of solidification frontiers exist [145, 146], that might be implemented in PRIME.

Last but not least, experimental validation of the above results is absolutely important, but has not been accomplished yet. From my point of view, the manipulation of metal foam with electromagnetic fields is very promising for the invention and roll out of new metal foam materials. I hope, I will see it coming into life one day, because, as the physicist Max Steenbeck said:

"For success, one needs more than the happiness of a subjective interest in his doing. One needs to be obsessed with the conviction of the objective importance of his work".

Bibliography

[1] I. Cantat, S. Cohen-Addad, F. Elias, F. Graner, R. Höhler, O. Pitois, Foams: structure and dynamics, Oxford University Press, 2013.

[2] H. Princen, A. Kiss, Osmotic pressure of foams and highly concentrated emulsions. 2. determination from the variation in volume fraction with height in an equilibrated column, Langmuir 3 (1987) 36–41.

[3] R. Höhler, Y. Y. C. Sang, E. Lorenceau, S. Cohen-Addad, Osmotic pressure and structures of monodisperse ordered foam, Langmuir 24 (2008) 418–425.

[4] L. Saulnier, L. Champougny, G. Bastien, F. Restagno, D. Langevin, E. Rio, A study of generation and rupture of soap films, Soft Matter (2014).

[5] E. Rio, W. Drenckhan, A. Salonen, D. Langevin, Unusually stable liquid foams, Advances in Colloid and Interface Science (2013).

[6] Pohltec Metalfoam GmbH, Metalfoam - Stabilität vereint Leichtigkeit De/Fr 05/2011,

[7] T. Hipke, Metal foam, Technical Report, Fraunhofer IWU, 2011.

[8] N. Dukhan, Metal Foams: Fundamentals and Applications, DEStech Publications, INC, 2013.

[9] M. F. Ashby, Metal foams: a design guide, Butterworth-Heinemann, 2000.

[10] L.-P. Lefebvre, J. Banhart, D. Dunand, Porous metals and metallic foams: current status and recent developments, Advanced Engineering Materials 10 (2008) 775–787.

[11] J. Banhart, Manufacturing routes for metallic foams, Journal of Metals 52 (2000) 22–27.

[12] J. Banhart, Manufacture, characterisation and application of cellular metals and metal foams, Progress in Materials Science 46 (2001) 559–632.

[13] J. Banhart, Metal foams: production and stability, Advanced Engineering Materials 8 (2006) 781–794.

[14] T. Wübben, J. Banhart, S. Odenbach, Production of metallic foam under low gravity conditions during parabolic flights, Microgravity Science and Technology 13 (2002) 36–42.

[15] N. Babcsán, F. Garcia-Moreno, D. Leitlmeier, J. Banhart, Liquid-metal foams–feasible in situ experiments under low gravity, in: Materials Science Forum, Trans. Tech. Publ., 2006.

[16] F. García-Moreno, C. Jiménez, M. Mukherjee, P. Holm, J. Weise, J. Banhart, Experiments on metallic foams under gravity and microgravity, Colloids and Surfaces A: Physicochemical and Engineering Aspects 344 (2009) 101–106.

[17] F. Garcia-Moreno, M. Mukherjee, C. Jimenez, J. Banhart, X-ray radioscopy of liquid metal foams under microgravity, Transactions of the Indian Institute of Metals 62 (2009) 451–454.

[18] O. Brunke, Untersuchung der Strukturentwicklung von Aluminiumschäumen mit Röntgenmethoden, Fortschritt-Berichte VDI: Reihe 3, Verfahrenstechnik, VDI-Verlag, 2006.

[19] O. Brunke, S. Odenbach, F. Beckmann, Quantitative methods for the analysis of synchrotron-μCT datasets of metallic foams, The European Physical Journal - Applied Physics 29 (2005) 73–81.

[20] O. Brunke, S. Odenbach, In situ observation and numerical calculations of the evolution of metallic foams, Journal of Physics - Condensed Matter 18 (2006) 6493–6506.

[21] P. A. Davidson, An introduction to magnetohydrodynamics,Cambridge University Press, 2001.

[22] T. Weier, G. Mutschke, G. Gerbeth, A. Alemany, A. Pilaud, On stability of the mhd flow around a cylinder in an aligned magnetic field, Magnetohydrodynamics 33 (1997) 11–18.

[23] G. Mutschke, V. Shatrov, G. Gerbeth, Cylinder wake control by magnetic fields in liquid metal flows, Experimental Thermal and Fluid Science 16 (1998) 92–99.

[24] P. M. Blosseville, S. Aleksandrova, S. Molokov, Buoyancy-driven mhd flow in electrically insulated rectangular ducts, Magnetohydrodynamics 43 (2007) 315–321.

[25] S. Schwarz, J. Fröhlich, Numerical simulation of mhd flow around fixed and freely ascending spheres and ellipsoids., in: 8th PAMIR International Conference on Fundamental and Applied MHD, 2011.

[26] T. Boeck, D. Krasnov, E. Zienicke, Numerical study of turbulent magnetohydrodynamic channel flow, Journal of Fluid Mechanics 572 (2007) 179–188.

[27] D. Leenov, A. Kolin, Theory of electromagnetophoresis. 1. Magnetohydrodynamic forces experienced by spherical and symmetrically oriented cylindrical particles, Journal of Chemical Physics 22 (1954) 683–688.

[28] A. van der Net, L. Blondel, A. Saugey, W. Drenckhan, Simulating and interpretating images of foams with computational ray-tracing techniques, Colloids and Surfaces A: Physicochemical and Engineering Aspects 309 (2007) 159–176.

[29] A. Maestro, W. Drenckhan, E. Rio, R. Höhler, Liquid dispersions under gravity: volume fraction profile and osmotic pressure, Soft Matter 9 (2013) 2531–2540.

[30] N. D. Denkov, V. Subramanian, D. Gurovich, A. Lips, Wall slip and viscous dissipation in sheared foams: Effect of surface mobility, Colloids and Surfaces A: Physicochemical and Engineering Aspects 263 (2005) 129–145.

[31] N. D. Denkov, S. Tcholakova, K. Golemanov, V. Subramanian, A. Lips, Foam–wall friction: Effect of air volume fraction for tangentially immobile bubble surface, Colloids and Surfaces A: Physicochemical and Engineering Aspects 282 (2006) 329–347.

[32] S. Cohen-Addad, R. Höhler, O. Pitois, Flow in foams and flowing foams, Annual Review of Fluid Mechanics 45 (2013) 241.

[33] R. Woodward, Surface Tension Measurements Using the Drop Shape Method., Technical Report, First Ten Angstroms, 2012.

[34] C. Lemaire, D. Langevin, Longitudinal surface waves at liquid interfaces: measurement of monolayer viscoelasticity, Colloids and Surfaces 65 (1992) 101–112.

[35] J. Banhart, H. Stanzick, L. Helfen, T. Baumbach, Metal foam evolution studied by synchrotron radioscopy, Applied Physics Letters 78 (2001) 1152–1154.

[36] L. Helfen, T. Baumbach, H. Stanzick, J. Banhart, A. Elmoutaouakkil, P. Cloetens, Viewing the early stages of metal foam formation by computed tomography using synchrotron radiation, Adavnced Engineering Materials 4(10) (2002) 808–813.

[37] Y.-C. Lee, K. J. Stebe, H.-S. Liu, S.-Y. Lin, Adsorption and desorption kinetics of $C_m E_8$ on impulsively expanded or compressed air–water interfaces, Colloids and Surfaces A: Physicochemical and Engineering Aspects 220 (2003) 139–150.

[38] Q. Song, A. Couzis, P. Somasundaran, C. Maldarelli, A transport model for the adsorption of surfactant from micelle solutions onto a clean air/water interface in the limit of rapid aggregate disassembly relative to diffusion and supporting dynamic tension experiments, Colloids and Surfaces A: Physicochemical and Engineering Aspects 282 (2006) 162–182.

[39] V. Fainerman, J. Petkov, R. Miller, Surface dilational viscoelasticity of $C_{14}EO_8$ micellar solution studied by bubble profile analysis tensiometry, Langmuir 24 (2008) 6447–6452.

[40] S. Cohen-Addad, R. Höhler, Y. Khidas, Origin of the slow linear viscoelastic response of aqueous foams, Physical Review Letters 93 (2004) 028302.

[41] N. D. Denkov, S. S. Tcholakova, R. Höhler, S. Cohen-Addad, Foam rheology, Foam Engineering: Fundamentals and Applications (2012) 91–120.

[42] S. Costa, R. Höhler, S. Cohen-Addad, The coupling between foam viscoelasticity and interfacial rheology, Soft Matter 9 (2013) 1100–1112.

[43] V. Bergeron, C. Radke, Equilibrium measurements of oscillatory disjoining pressures in aqueous foam films, Langmuir 8 (1992) 3020–3026.

[44] V. Bergeron, Disjoining pressures and film stability of alkyltrimethylammonium bromide foam films, Langmuir 13 (1997) 3474–3482.

[45] A. T. Dinsdale, P. N. Quested, The viscosity of aluminium and its alloys - a review of data and models, Journal of Materials Science 39 (2004) 7221–7228.

[46] M. J. Assael, K. Kakosimos, R. M. Banish, J. Brillo, I. Egry, R. Brooks, P. N. Quested, K. C. Mills, A. Nagashima, Y. Sato, et al., Reference data for the density and viscosity of liquid aluminum and liquid iron, Journal of Physical and Chemical Reference Data 35 (2006) 285–300.

[47] A. Haibel, A. Rack, J. Banhart, Why are metal foams stable?, Applied Physics Letters 89 (2006) 154102.

[48] R. Clift, J. R. Grace, M. E. Weber, Bubbles, Drops and Particles, Dover Publications Inc., 1978.

[49] T. Kempe, S. Schwarz, J. Fröhlich, Modelling of spheroidal particles in viscous flow., in: Academy Colloquium Immersed Boundary Methods: Current Status and Future Research Directions, Amsterdam, the Netherlands, 2009.

[50] T. Kempe, J. Fröhlich, On euler-lagrange coupling and collision modelling for spherical particles, in: 8th International ERCOFTAC Symposium on Engineering Turbulence Modelling and Measurements, Marseille, France, 2010.

[51] T. Kempe, J. Fröhlich, Collision modeling for spherical particles in viscous fluids, in: 7th International Conference on Multiphase Flow, Tampa, Florida, USA, 2010.

[52] T. Kempe, A numerical method for interface-resolving simulations of particle-laden flows with collisions, TUDpress Verlag der Wissenschaften GmbH, 2013.

[53] T. Kempe, J. Fröhlich, An improved immersed boundary method with direct forcing for the simulation of particle laden flows, Journal of Computational Physics 231 (2012) 3663–3684.

[54] S. Schwarz, J. Fröhlich, Simulation of a bubble chain in a container of high aspect ratio exposed to a magnetic field, The European Physical Journal Special Topics 220 (2013) 195–205.

[55] B. Vowinckel, J. Fröhlich, Simulation of bed load transport in turbulent open channel flow, PAMM 12 (2012) 505–506.

[56] S. Schwarz, J. Fröhlich, DNS of single bubble motion in liquid metal and the influence of a magnetic field., in: 7th International Symposium on Turbulence and Shear Flow Phenomena, Ottawa, Kanada, 2011.

[57] F. Durst, Grundlagen der Strömungsmechanik, Springer, 2006.

[58] J. H. Ferziger, M. Perić, Computational methods for fluid dynamics, Springer Berlin, 1996.

[59] C. S. Peskin, Numerical analysis of blood flow in the heart, Journal of Computational Physics 25 (1977) 220–252.

[60] M. Uhlmann, An immersed boundary method with direct forcing for the simulation of particulate flows, Journal of Computational Physics 209 (2005) 448–476.

[61] S. Heitkam, S. Schwarz, J. Fröhlich, Simulation of the influence of electromagnetic fields on the drainage in wet metal foam, Magnetohydrodynamics 48 (2012) 313–320.

[62] S. Schwarz, An immersed boundary method for particles and bubbles in magnetohydrodynamic flows, Dissertation, TU Dresden, 2013.

[63] D. Durian, Foam mechanics at the bubble scale, Physical Review Letters 75 (1995) 4780.

[64] C. S. O'Hern, L. E. Silbert, A. J. Liu, S. R. Nagel, Jamming at zero temperature and zero applied stress: The epitome of disorder, Physical Review E 68 (2003) 011306.

[65] K. Johnson, K. Kendall, A. Roberts, Surface energy and the contact of elastic solids, Proceedings of the royal society of London. A. Mathematical and Physical Sciences 324 (1971) 301–313.

[66] K. L. Johnson, K. L. Johnson, Contact mechanics, Cambridge University Press, 1987.

[67] H. Brenner, The slow motion of a sphere through a viscous fluid towards a plane surface, Chemical Engineering Science 16 (1961) 242–251.

[68] R. G. Cox, H. Brenner, The slow motion of a sphere through a viscous fluid towards a plane surface - small gap widths, including inertial effects, Chemical Engineering Science 22 (1967) 1753–1777.

[69] A. Goldman, R. G. Cox, H. Brenner, Slow viscous motion of a sphere parallel to a plane wall. 1. Motion through a quiescent fluid, Chemical Engineering Science 22 (1967) 637–651.

[70] R. H. Davis, J.-M. Serayssol, E. Hinch, The elastohydrodynamic collision of two spheres, Journal of Fluid Mechanics 163 (1986) 479–497.

[71] T. Kempe, J. Fröhlich, Collision modeling for the interface-resolved simulation of spherical particles in viscous fluids, Journal of Fluid Mechanics 709 (2012) 445–489.

[72] T. Kempe, B. Vowinckel, J. Fröhlich, On the relevance of collision modeling for interface-resolving simulations of sediment transport in open channel flow, International Journal of Multiphase Flow 58 (2014) 214–235.

[73] M. H. Hendrix, R. Manica, E. Klaseboer, D. Y. Chan, C.-D. Ohl, Spatiotemporal evolution of thin liquid films during impact of water bubbles on glass on a micrometer to nanometer scale, Physical Review Letters 108 (2012) 247803.

[74] R. R. Dagastine, R. Manica, S. L. Carnie, D. Chan, G. W. Stevens, F. Grieser, Dynamic forces between two deformable oil droplets in water, Science 313 (2006) 210–213.

[75] D. Y. Chan, E. Klaseboer, R. Manica, Film drainage and coalescence between deformable drops and bubbles, Soft Matter 7 (2011) 2235–2264.

[76] D. Y. Chan, E. Klaseboer, R. Manica, Theory of non-equilibrium force measurements involving deformable drops and bubbles, Advances in Colloid and Interface Science 165 (2011) 70–90.

[77] R. Manica, E. Klaseboer, R. Gupta, M. H. Hendrix, C.-D. Ohl, D. Y. Chan, Modelling film drainage of a bubble hitting and bouncing off a surface, 9th International Conference on CFD in the Minerals and Process Industries 100 (2012) 024501.

[78] R. W. Johnson, Handbook of fluid dynamics, Crc Press, 1998.

[79] F. Morrison, M. Stewart, Small bubble motion in an accelerating liquid, Journal of Applied Mechanics 43 (1976) 399–403.

[80] G. Haller, T. Sapsis, Where do inertial particles go in fluid flows?, Physica D: Nonlinear Phenomena 237 (2008) 573–583.

[81] E. Abbena, S. Salamon, A. Gray, Modern differential geometry of curves and surfaces with Mathematica, CRC Press, 2006.

[82] R. Zenit, D. Legendre, The coefficient of restitution for air bubbles colliding against solid walls in viscous liquids, Physics of Fluids 21 (2009) 083306.

[83] J. Zawala, M. Krasowska, T. Dabros, K. Malysa, Influence of bubble kinetic energy on its bouncing during collisions with various interfaces, The Canadian Journal of Chemical Engineering 85 (2007) 669–678.

[84] F. Vincent, A. Le Goff, G. Lagubeau, D. Quéré, Bouncing bubbles, Journal of Adhesion 83 (2007) 897–906.

[85] P. Garstecki, M. J. Fuerstman, H. A. Stone, G. M. Whitesides, Formation of droplets and bubbles in a microfluidic T-junction: scaling and mechanism of break-up, Lab on a Chip 6 (2006) 437–446.

[86] J. Richardson, W. Zaki, The sedimentation of a suspension of uniform spheres under conditions of viscous flow, Chemical Engineering Science 3 (1954) 65–73.

[87] H. Nicolai, B. Herzhaft, E. J. Hinch, L. Oger, E. Guazzelli, Particle-velocity fluctuations and hydrodynamic self-diffusion of sedimenting non-brownian spheres, Physics of Fluids 7 (1995) 12–23.

[88] R. DiFelice, E. Parodi, Wall effects on the sedimentation velocity of suspensions in viscous flow, Aiche Journal 42 (1996) 927–931.

[89] P. N. Segre, E. Herbolzheimer, P. M. Chaikin, Long-range correlations in sedimentation, Physical Review Letters 79 (1997) 2574–2577.

[90] B. Hölting, W. G. Coldewey, Hydrogeologie: Einführung in die Allgemeine und Angewandte Hydrogeologie, Spektrum Akademischer Verlag, 2005.

[91] D. Weaire, N. Pittet, S. Hutzler, D. Pardal, Steady-state drainage of an aqueous foam, Physical Review Letters 71 (1993) 2670–2673.

[92] A. Hazen, Experiments upon the purification of sewage and water at the lawtence experiment station, Massachusetts States Board of Health Twenty-Third Annual Report (1892) 428–434.

[93] S. Heitkam, Y. Yoshitake, F. Toquet, D. Langevin, A. Salonen, Speeding up of sedimentation under confinement, Physical Review Letters 110 (2013) 178302.

[94] S. Heitkam, W. Drenckhan, J. Fröhlich, Packing spheres tightly: Influence of mechanical stability on close-packed sphere structures, Physical Review Letters 108 (2012) 148302.

[95] A. van der Net, W. Drenckhan, I. Weaire, S. Hutzler, The crystal structure of bubbles in the wet foam limit, Soft Matter 2 (2006) 129–134.

[96] A. van der Net, G. W. Delaney, W. Drenckhan, D. Weaire, S. Hutzler, Crystalline arrangements of microbubbles in monodisperse foams, Colloids and Surfaces A: Physicochemical and Engineering Aspects 309 (2007) 117–124.

[97] W. Drenckhan, D. Langevin, Monodisperse foams in one to three dimensions, Current Opinion in Colloid & Interface Science 15 (2010) 341–358.

[98] J. P. Hoogenboom, A. Yethiraj, A. K. van Langen-Suurling, J. Romijn, A. van Blaaderen, Epitaxial crystal growth of charged colloids, Physical Review Letters 89 (2002) 256104.

[99] F. Ramiro-Manzano, F. Meseguer, E. Bonet, I. Rodriguez, Faceting and commensurability in crystal structures of colloidal thin films, Physical Review Letters 97 (2006) 028304.

[100] H. Miguez, F. Meseguer, C. Lopez, A. Mifsud, J. S. Moya, L. Vazquez, Evidence of fcc crystallization of SiO_2 nanospheres, Langmuir 13 (1997) 6009–6011.

[101] A. B. Yu, X. Z. An, R. P. Zou, R. Y. Yang, K. Kendall, Self-assembly of particles for densest packing by mechanical vibration, Physical Review Letters 97 (2006) 265501.

[102] L. V. Woodcock, Entropy difference between the face-centred cubic and hexagonal close-packed crystal structures, Nature 385 (1997) 141–143.

[103] T. Titscher, S. Heitkam, C. D. Kreuter, W. Drenckhan, D. Hajnal, F. Piechon, J. Fröhlich, Elastic properties of material with spherical voids in different arrangements, Intended for: Journal of the Mechanics and Physics of Solids to be submitted.

[104] T. Titscher, Untersuchung der Manipulation mechanischer Eigenschaften von Metallschäumen über elektromagnetische Felder im Entstehungsprozess, Diplomarbeit, TU Dresden (2013) .

[105] F. Radjai, S. Roux, J. J. Moreau, Contact forces in a granular packing, Chaos: An Interdisciplinary Journal of Nonlinear Science 9 (1999) 544–550.

[106] C. S. O'Hern, S. A. Langer, A. J. Liu, S. R. Nagel, Force distributions near jamming and glass transitions, Physical Review Letters 86 (2001) 111–114.

[107] N. W. Mueggenburg, H. M. Jaeger, S. R. Nagel, Stress transmission through three-dimensional ordered granular arrays, Physical Review E 66 (2002) 031304.

[108] W. Sanders, L. Gibson, Mechanics of hollow sphere foams, Materials Science and Engineering: A 347 (2003) 70–85.

[109] A. Ngan, On distribution of contact forces in random granular packings, Physica A: Statistical Mechanics and its Applications 339 (2004) 207–227.

[110] J. Yin, M. Retsch, E. L. Thomas, M. C. Boyce, Collective mechanical behavior of multilayer colloidal arrays of hollow nanoparticles, Langmuir 28 (2012) 5580–5588.

[111] X. An, A. Yu, Analysis of the forces in ordered fcc packings with different orientations, Powder Technology 248 (2013) 121–130.

[112] R. Hill, A self-consistent mechanics of composite materials, Journal of the Mechanics and Physics of Solids 13 (1965) 213–222.

[113] B. Budiansky, On the elastic moduli of some heterogeneous materials, Journal of the Mechanics and Physics of Solids 13 (1965) 223–227.

[114] T. Iwakuma, S. Nemat-Nasser, Composites with periodic microstructure, Computers & Structures 16 (1983) 13–19.

[115] J. W. Eischen, S. Torquato, Determining elastic behavior of composites by the boundary element method, Journal of Applied Physics 74 (1993) 159–170.

[116] S. Torquato, Exact expression for the effective elastic tensor of disordered composites, Physical Review Letters 79 (1997) 681–684.

[117] S. Torquato, Effective stiffness tensor of composite media: 1. Applications to isotropic dispersions, Journal of the Mechanics and Physics of Solids 46 (1998) 1411–1440.

[118] R. M. Christensen, A critical evaluation for a class of micro-mechanics models, Journal of the Mechanics and Physics of Solids 38 (1990) 379–404.

[119] I. Cohen, Simple algebraic approximations for the effective elastic moduli of cubic arrays of spheres, Journal of the Mechanics and Physics of Solids 52 (2004) 2167–2183.

[120] A. Day, K. Snyder, E. Garboczi, M. Thorpe, The elastic moduli of a sheet containing circular holes, Journal of the Mechanics and Physics of Solids 40 (1992) 1031–1051.

[121] J. Segurado, J. Llorca, A numerical approximation to the elastic properties of sphere-reinforced composites, Journal of the Mechanics and Physics of Solids 50 (2002) 2107–2121.

[122] Y. Ni, M. Y. Chiang, Prediction of elastic properties of heterogeneous materials with complex microstructures, Journal of the Mechanics and Physics of Solids 55 (2007) 517–532.

[123] M. Bouhlel, M. Jamei, C. Geindreau, Microstructural effects on the overall poroelastic properties of saturated porous media, Modelling and Simulation in Materials Science and Engineering 18 (2010) 045009.

[124] M. Saadatfar, M. Mukherjee, M. Madadi, G. Schröder-Turk, F. Garcia-Moreno, F. Schaller, S. Hutzler, A. Sheppard, J. Banhart, U. Ramamurty, Structure and deformation correlation of closed-cell aluminium foam subject to uniaxial compression, Acta Materialia 60 (2012) 3604–3615.

[125] M. Skoge, A. Donev, F. H. Stillinger, S. Torquato, Packing hyperspheres in high-dimensional euclidean spaces, Physical Review E 74 (2006) 041127.

[126] W. Drugan, J. Willis, A micromechanics-based nonlocal constitutive equation and estimates of representative volume element size for elastic composites, Journal of the Mechanics and Physics of Solids 44 (1996) 497–524.

[127] W. Drugan, Micromechanics-based variational estimates for a higher-order nonlocal constitutive equation and optimal choice of effective moduli for elastic composites, Journal of the Mechanics and Physics of Solids 48 (2000) 1359–1387.

[128] D. Kreuter, K. Sepahvand, M. Beitelschmidt, Homogenization with uncertain input data, PAMM 11 (2011) 533–534.

[129] G. A. Holzapfel, Nonlinear solid mechanics, volume 24, Wiley Chichester, 2000.

[130] W. Voigt, Über die Beziehung zwischen den beiden Elastizitätskonstanten isotroper Körper, Annalen der Physik 274 (1889) 573–587.

[131] G. Verbist, D. Weaire, A. Kraynik, The foam drainage equation, Journal of Physics: Condensed Matter 8 (1996) 3715–3731.

[132] S. Cox, D. Weaire, S. Hutzler, J. Murphy, R. Phelan, G. Verbist, Applications and generalizations of the foam drainage equation, Proceedings of the Royal Society of London. Series A: Mathematical, Physical and Engineering Sciences 456 (2000) 2441–2464.

[133] A. Sellier, Migration of a conducting ellipsoid subject to uniform ambient electric and magnetic fields, Comptes Rendus Mecanique 331 (2003) 127–132.

[134] A. Sellier, Motion of arbitrary 2-bubble clusters immersed in a conducting liquid when subject to uniform ambient electric and magnetic fields, Magnetohydrodynamics 48 (2012) 137–145.

[135] R. B. Kaplan, Open cell tantalum structures for cancellous bone implants and cell and tissue receptors, 1994. US Patent 5,282,861.

[136] C. Wen, Y. Yamada, K. Shimojima, Y. Chino, H. Hosokawa, M. Mabuchi, Novel titanium foam for bone tissue engineering, Journal of Materials Research 17 (2002) 2633–2639.

[137] E. D. Spoerke, N. G. Murray, H. Li, L. C. Brinson, D. C. Dunand, S. I. Stupp, A bioactive titanium foam scaffold for bone repair, Acta Biomaterialia 1 (2005) 523–533.

[138] K. A. Brakke, The surface evolver, Experimental mathematics 1 (1992) 141–165.

[139] R. Phelan, D. Weaire, K. Brakke, Computation of equilibrium foam structures using the surface evolver, Experimental Mathematics 4 (1995) 181–192.

[140] K. A. Brakke, The surface evolver and the stability of liquid surfaces, Philosophical Transactions of the Royal Society of London. Series A: Mathematical, Physical and Engineering Sciences 354 (1996) 2143–2157.

[141] S. Cox, A viscous froth model for dry foams in the surface evolver, Colloids and Surfaces A: Physicochemical and Engineering Aspects 263 (2005) 81–89.

[142] F. Graner, Y. Jiang, E. Janiaud, C. Flament, Equilibrium states and ground state of two-dimensional fluid foams, Physical Review E 63 (2000) 011402.

[143] K. Götze, personal communication, HZDR Dresden, 2014.

[144] V. M. Schmidt, Elektrochemische Verfahrenstechnik: Grundlagen, Reaktionstechnik, Prozessoptimierung, John Wiley & Sons, 2012.

[145] M. Selder, L. Kadinski, Y. Makarov, F. Durst, P. Wellmann, T. Straubinger, D. Hofmann, S. Karpov, M. Ramm, Global numerical simulation of heat and mass transfer for SiC bulk crystal growth by PVT, Journal of Crystal Growth 211 (2000) 333–338.

[146] I. Steinbach, M. Apel, T. Rettelbach, D. Franke, Numerical simulations for silicon crystallization processes: examples from ingot and ribbon casting, Solar Energy Materials and Solar Cells 72 (2002) 59–68.

11 Appendix

11.1 Symbols

Symbol	Unit	Description
α_{el}	–	Phase indicator for electric conductivity
δ	m	Distance between sphere and obstacle during collision
Δ	m	Distance deficit
Δ_0	m	Minimal distance deficit for collision model
ε	–	Strain tensor
μ_0	$^{\mathrm{Vs}}/_{\mathrm{Am}}$	Magnetic constant
μ_{f}	$5\,^{\mathrm{kg}}/_{\mathrm{sm}}$	Dynamic viscosity of the fluid
ν_{f}	$^{\mathrm{m}^2}/_{\mathrm{s}}$	Kinematic viscosity of the fluid
ν_{m}	–	Poisson ratio of the solid material
ρ_{b}	$^{\mathrm{kg}}/_{\mathrm{m}^3}$	Bubble density
ρ_{f}	$^{\mathrm{kg}}/_{\mathrm{m}^3}$	Fluid density
σ	$^{\mathrm{N}}/_{\mathrm{m}}$	Surface tension
σ	$^{\mathrm{N}}/_{\mathrm{m}^2}$	Stress tensor
σ_{el}	$^{\mathrm{S}}/_{\mathrm{m}}$	Electrical conductivity
Φ	–	Gas fraction
Φ_{el}	V	Electric potential
$\Phi_{\mathrm{el,cor}}$	V	Correction of the electric potential
Φ_{s}	–	Solid fraction
Ω_{b}	m^2	Bubble surface
A_{contact}	m^2	Contact area during collision
$\mathbf{B}$	T	Magnetic field strength
B_0	T	Applied magnetic field strength
c	$^{\mathrm{m}}/_{\mathrm{s}}$	Speed of light
$\mathbf{C}$	$^{\mathrm{N}}/_{\mathrm{m}^2}$	Stiffness matrix
C_1	$^{\mathrm{m}}/_{\mathrm{s}}$	Fitting constant
C_2	$^{\mathrm{m}}/_{\mathrm{sT}^2}$	Fitting constant
C_{am}	–	Added mass coefficient

Symbol	Unit	Description
C_{bc}	–	Parameter for boundary condition in viscous collision force
C_{coll}	N/m	Collision spring constant
c_{D}	–	Drag coefficient (of a sphere)
D	m	Bubble diameter
$\mathbf{D}$	m^2/N	Compliance matrix
E	N/m^2	Young's modulus
$\mathbf{E}$	m/s	Electric field
E_0	N/m^2	Young's modulus of the solid material
E_{el}	Nm	Elastic energy
E_i	N/m^2	Young's modulus for deformation in i-direction
$E_{\text{kin,b}}$	Nm	Kinetic energy of a rising bubble
E_{m}	N/m^2	Average Young's modulus
ΔE_{max}	Nm	Maximum energy dissipation in a lamella
Eo	–	Eötvös number
$\mathbf{e}_{\text{n}}$	–	Unit vector normal to a certain plane
E_{s}	Nm	Energy of the surface area of a bubble
$\mathbf{e}_{\text{t}}$	–	Unit vector tangential to a certain plane
E_{Voigt}	N/m^2	Young's modulus according to Voigts rule of mixture
$\mathbf{e}_x$, $\mathbf{e}_y$, $\mathbf{e}_z$	–	Unit vectors in x, y and z direction
$\mathbf{f}$	N/m^3	Volumetric force
F_{a}	N	Force acting on the bubble from contact area
F_{acc}	N	Tip force for testing the packing stability
F_{b}	N	Buoyancy force of a bubble
$\mathbf{F}_{\text{coll}}$	N	Collision force
F_{D}	N	Drag force of a rising bubble (sphere)
F_{elastic}	N	Normal elastic collision force
$\mathbf{F}_{\text{fluid}}$	N	Fluid force, acting on a bubble
$\mathbf{f}_{\text{IBM}}$	N/m^3	Coupling forces in the IBM approach
$\mathbf{f}_{\text{L}}$	N/m^3	Lorentz force density
F_{out}	N	Force acting on the bubble from external pressure field
F_{tang}	N	Tangential viscous collision force
F_{visc}	N	Normal viscous collision force
g	m/s^2	Gravitational acceleration
g_{L}	m/s^2	Acceleration resulting from the electromagnetic field
h_0	m	Thickness of lamella during collision
H_{m}	$1/\text{m}$	Mean curvature of a surface
H_{packing}	m	Height of the Bubble agglomeration
i	–	Counter

Symbol	Unit	Description
$\mathbf{j}$	$\mathrm{A/m^2}$	Electric current density
j_0	$\mathrm{A/m^2}$	Applied electric current
$\mathbf{K}$	–	Transformation matrix
k_f	$\mathrm{m/s}$	Hydraulic permeability
l_l	m	Thickness of a lamella in crystalline arranged spheres
L_l	m	Lattice spacing in crystalline arranged spheres
L_ref	m	Reference length
L_x, L_y, L_z	m	Size of the computational domain in x, y and z direction
m_am	kg	Added mass of a bubble [78]
m_b	kg	Bubble mass
n_b	–	Number of bubbles
n_cryst	–	Number of bubbles in crystalline arrangement
n_FCC, n_HCP	–	Number of bubbles in FCC and HCP arrangement
N_g	–	Number of grid points per bubble diameter
n_L	–	Number of Lagrangian marker points
p_∞	Pa	Pressure far away
Δp_0	$\mathrm{N/m^2}$	Pressure drop at zero magnetic field strength
p_b	Pa	Overpressure inside a bubble
P_FCC	–	Quantity for the preference of FCC over HCP packing
p_gap	Pa	Pressure in the gap between bubble and obstacle
p_out	Pa	Fluid pressure next to the bubble surface
Δp	$\mathrm{m^2/s^2}$	Steady pressure drop in static drainage
$\mathbf{r}$	m	Position vector
R_a	m	Radius of the circular contact area during collision
R_b	m	Curvature radius of the surface in the dissipation zone during collision
Re	–	Reynolds number
R_eq	m	Radius of a sphere with the same volume as the (deformed) bubble
R_m	–	Magnetic Reynolds number
t	s	Time
Δs	m	Numerical grid step
s_s	m	Step width for shape calculation
Δt	s	Numerical time step
t_ref	s	Reference time

Symbol	Unit	Description
$\mathbf{u}$	m/s	Fluid velocity
$\mathbf{u}_\mathrm{b}$	m/s	Bubble velocity
u_c	m/s	Velocity of a bubble, upward or normal to an obstacle
v_0	m/s	Average downward flow velocity in static drainage
V	m^3	Volume
$\dot{V}_0$	m^3/s	Volumetric flow in static drainage
V_b	m^3	Bubble volume
v_ref	m/s	Undisturbed rising velocity of a single bubble
x, y, z	m	Components of the position vector
$\mathbf{x}_\mathrm{c}$	m	Centre position of a bubble
Δy_c	m	Distance between two consecutive layers of mobile bubbles

Abbreviation	Meaning
BCC	Body-centred cubic
DNS	Direct numerical simulation
EHC	Equivalent homogeneous continuum
FCC	Face-centred cubic
FCCh	Face-centred cubic (hexagonally orientated)
FEM	Finite-Element method
HCP	Hexagonally close-packed
IBM	Immersed-Boundary method
MHD	Magnetohydrodynamic
PRIME	Phase-resolving simulation environment
RCP	Random close-packed
RHCP	Random hexagonally close-packed
RVE	Representative volume element
SC	Simple cubic

11.2   The author's publications

11.2.1   Peer reviewed articles

S. Heitkam, S. Schwarz, C. Santarelli, J. Fröhlich. Influence of an electromagnetic field on the formation of wet metal foam. *Eur. Phys. J. Special Topics*, Vol. 220, 207-214, 2013

J. Fröhlich, S. Schwarz, **S. Heitkam**, C. Santarelli, C. Zhang, T. Vogt, S. Boden, A. Andruszkiewicz, K. Eckert, S. Odenbach, S. Eckert. Influence of magnetic fields on the behavior of bubbles in liquid metals. *Eur. Phys. J. Special Topics*, Vol. 220, 167-183, 2013

S. Heitkam, Y. Yoshitake, F. Toquet, D. Langevin, A. Salonen. Speeding up of sedimentation under confinement. *Phys.Rev.Lett.*, Vol. 110, 178302, 2013

S. Heitkam, S. Schwarz, J. Fröhlich. Simulation of the influence of electromagnetic fields on the drainage in wet metal foam. *Magnetohydrodyn.*, Vol. 48 (2), 313-320, 2012,

S. Heitkam, W. Drenckhan, J. Fröhlich. Packing spheres tightly: Influence of mechanical stability on close-packed sphere structures. *Phys.Rev.Lett.*, Vol. 108 (14), 148302, 2012
discussed in Physikjounal 06/2012 "Aus Schaum gebaut"

S. Heitkam, J. Fröhlich. Formation of crystalline bubble structure in wet metal foams. *Proc.Appl.Math.Mech*, Vol. 11, 653-654, 2011

A. Voigt, **S. Heitkam**, L. Büttner, J. Czarske. A Bessel beam laser Doppler velocimeter. *Optics Communications*, 2009

11.2.2   Articles in preparation

S. Heitkam, A. Sommer, W. Drenckhan, J. Fröhlich. A simple collision model for small bubbles. *intended for Soft Matter*, **to be submitted**

T. Titscher, **S. Heitkam**, D. Kreuter, W. Drenckhan, D. Hajnal, F. Piechon, J. Fröhlich. Elastic properties of material with spherical voids in different arrangements. *intended for J. Mech. Phys. Solides*, **to be submitted**

11.2.3 Conference contributions (presenting author*)

S. Heitkam, P. Yazhgur, Y. Yoshitake, F. Toquet, D. Langevin, J. Lintuvuori, H. H. Wensink, G. Foffi, A. Salonen*. *Speeding up of sedimentation under Confinement.* Liquid Matter Conference, Lisbon, Portugal, July 21-25, 2014

W. Drenckhan*, M. Roché, L. Champougny, T. Gaillard, **S. Heitkam**, F. Piechon, C. Poulard, E. Rio. *A science of transition: from liquid to solid foams.* 10th EU-FOAM, Thessaloniki, Greece, July 7-10, 2014

P.Yazhgur*, Y. He, **S. Heitkam**, D. Langevin, A. Salonen. *Adsorption/desorption dynamics of surfactants studied by bubble compression method.* 20th International Symposium on Surfactants in Solution. Coimbra, Portugal, June 22-27, 2014

S. Heitkam*, A. Sellier, S. Schwarz, J. Fröhlich. *Action on an insolating sphere moving close to a wall - a comparative study.* 9th PAMIR International Conference on Fundamental and Applied MHD. Riga, Latvia, June 16-20, 2014

S. Heitkam*, H. Ihlenfeld, J. Fröhlich. *Investigation of potential flow using the soap film analogy - a historical review.* 85th Annual Meeting of the International Association of Applied Mathematics and Mechanics, Erlangen, Germany, March 10-14, 2014

S. Heitkam*, A. Sommer, J. Fröhlich. *Collision modeling for small bubbles.* 11th Multiphase Flow Conference, Rossendorf, Germany, Nov. 26-28, 2013

S. Heitkam, P. Yazhgur, Y. Yoshitake, F. Toquet, D. Langevin, A. Salonen*. *Speeding up of sedimentation under Confinement.* International Soft Matter Conference, Rome, Italy, September 15-19, 2013

S. Heitkam*, S. Schwarz, C. Santarelli, J. Fröhlich. *Artificial zero-gravity in wet metal foam by means of electromagnetic fields.* 8th ICMF International Conference on Multiphase Flow, Jeju, Korea, May 26-31, 2013.

S. Heitkam*, S. Schwarz, J. Fröhlich. *Manipulation of metal foam with electromagnetic fields.* Int. Symp. Electromagnetic Flow Control, Dresden, Germany, Jan. 16-18, 2013.

S. Heitkam*, J. Fröhlich, W. Drenckhan. *Why bubble crystals prefer fcc over hcp packing.* 9th EUFOAM, Lisboa, Portugal, July 8-11, 2012

S. Heitkam*, A. Hoffmann, W. Drenckhan, D. Langevin, J. Fröhlich. *Numerical simulation of bubble collisions with PRIME.* CECAM Dissipative rheology of foam, Dublin, Ireland, January 9-12, 2012

S. Heitkam*, S. Schwarz, J. Fröhlich. *Simulation of the influence of electromag-*

netic fields on the drainage in wet metal foam. 8th PAMIR International Conference on Fundamental and Applied MHD, Borgo, France, September 5-9, 2011

S. Heitkam*, J. Fröhlich. *Formation of crystalline bubble structure in wet metal foam.* 82th Annual Meeting of the International Association of Applied Mathematics and Mechanics, Graz, Austria, April 18-21, 2011

M. Neumann, K. Shirai, **S. Heitkam***, L. Büttner, J. Czarske. *Measurement of two-point velocity correlations in turbulent free shear flows with extended Laser Doppler Velocity profile sensor.* 17. Fachtagung Lasermethoden in der Strömungsmesstechnik, Erlangen, 8. - 10. September 2009

11.2.4   Invited lectures

S. Heitkam*, J. Fröhlich, W. Drenckhan. *Why do spheres prefer fcc packing when subjected to external forces.* International Workshop on PACKING PROBLEMS, Dublin, Ireland, September 2-5, 2012

11.3 Student theses

Under the author's guidance, several students carried out their thesis.

A. Renz: *Messung der partikelbeladenen Umströmung von Hochgeschwindigkeitszügen.* Diplomarbeit. (2010)

S. Ray: *Analyse von Partikeltrajektorien aus Phasenaufgelösten Simulationen von Sediment.* Schulprojekt. (2011)

A. Hoffmann: *Auswahl und Validierung eines geeigneten Kollisionsmodells für Blasen.* Großer Beleg. (2011)

R. Hoyer: *Aufbau und Charakterisierung eines Lased-Doppler Anemometers.* Großer Beleg. (2011)

F. Toquet: *Investigation of creaming and sedimentation in small capillaries.* Master thesis. (2012)

T. Titscher: *Untersuchung der Manipulation mechanischer Eigenschaften von Metallschäumen über elektromagnetische Felder im Entstehungsprozess.* Diplomarbeit. (2013)

J. Amslinger *Untersuchung der magnetohydrodynamischen Kräfte auf eine Blase in Wandnähe.* Großer Beleg. (2014)

CURRICULUM VITAE   November 2014

Professur für Strömungsmechanik
Technische Universität Dresden
George-Bähr-Str. 3c, room ZEU 153
D-01062 Dresden

Born: 18.12.1984
German nationality
Married, 2 children (*2012, 2014)

Tel: +49 (0)351 463 34910
Email: sascha.heitkam@tu-dresden.de

EDUCATION

2010 - 2014	**Doctoral degree,** *summa cum laude* at TU Dresden, Germany and Université Paris-Sud XI, France. Simulation of the generation of metal foam with electromagnetic fields.
2007 - 2009	**Diploma in Power Engineering** with grade 1.1, best of class (out of approx. 800), TU Dresden
2005 - 2007	**Preliminary diploma in Physics,** with grade 1.7, TU Dresden **Preliminary diploma in Mechanical Engineering,** with grade 1.6, TU Dresden
1997 - 2004	**Abitur** with grade 1.0, Max-Steenbeck-Gymnasium, Cottbus, Germany

MAJOR GRANTS, FELLOWSHIPS AND AWARDS

2011 - 2012	Eiffel Scholarship for research at the Université Paris-Sud XI, France
2010	Enno-Heidebroek-Award for excellence in the diploma studies
2010	Festo Award for best graduation of class
2008	Fellowship of Vattenfall (1 year)
2004	Young Scientists Contest: 2nd prize (Germany) *acoustics of foam*
2000 - 2004	Various prizes in Physics-, Mathematics-, Chemistry- and Young Scientists Contests.

WORK EXPERIENCE

2010 - 2014	**Doctoral studies** Chair of Fluid Mechanics, TU Dresden, Germany and Laboratoire de Physique des Solides, Université Paris-Sud XI, France
2009 (6 Mo)	**Student placement** ABB Turbosystems, Baden, Switzerland
2007-2009	**Student assistant** Chair of Measurement and Testing Technique, TU Dresden
2006 (2 Mo)	**Student placement** at AMD Saxony, Dresden, Germany
2005 (2 Mo)	**Student placement** at Vestas Blades Germany, AG, Lauchhammer

www.ingramcontent.com/pod-product-compliance
Lightning Source LLC
Chambersburg PA
CBHW081538220326
41598CB00036B/6477